WORKBOOKS IN CHEMISTRY

SERIES EDITOR

STEPHEN K. SCOTT

WORKBOOKS IN CHEMISTRY

Beginning mathematics for chemistry

Stephen K. Scott

WORKBOOKS IN CHEMISTRY

Beginning Mathematics for Chemistry

STEPHEN K. SCOTT

School of Chemistry
University of Leeds

OXFORD   NEW YORK   TOKYO
OXFORD UNIVERSITY PRESS
1995

Oxford University Press, Walton Street, Oxford OX2 6DP
Oxford New York
Athens Auckland Bangkok Bombay
Calcutta Cape Town Dar es Salaam Delhi
Florence Hong Kong Istanbul Karachi
Kuala Lumpur Madras Madrid Melbourne
Mexico City Nairobi Paris Singapore
Taipei Tokyo Toronto
and associated companies in
Berlin Ibadan

Oxford is a trade mark of Oxford University Press

Published in the United States
by Oxford University Press Inc., New York

A catalogue record for this book is available from the British Library

Library of Congress Cataloging in Publication Data
(Data applied for)

ISBN 0 19 855930 5

Typeset by the author
Printed in Great Britain by
Progressive Printing UK Ltd
Leigh-on-Sea, Essex

Workbooks in Chemistry: Series Preface

The new *Workbooks in Chemistry Series* is designed to provide support to students in their learning in areas that cannot be covered in great detail in formal courses. The format allows individual, self-paced study. Students can also work in groups guided by tutors. Teaching staff can monitor progress as the students complete the exercises in the text. The Workbooks aim to support the more traditional teaching methods such as lectures. The format of the Workbooks has been evolved through experience and discussions with students over several years. Students benefit through the Examples and Exercises that provide practice and build confidence. University staff faced with increasing class sizes may find the Workbooks helpful in encouraging 'self-learning' and meeting the individual needs of their students more efficiently. The topics covered in the early Workbooks in the Series will concentrate on background support appropriate to the early years of a Chemistry degree, including mathematics, performing calculations, and basic concepts in organic chemistry. These should also be of interest to students who are taking chemistry courses as part of other degree schemes, such as biochemistry and environmental sciences. Later Workbooks will be designed to support material typically encountered in later years of a Chemistry course.

Preface

This workbook has grown out of an examples-class approach to teaching the limited range of mathematical skills needed by chemists, to students entering the School of Chemistry at Leeds who do not have appropriate preparation at A-level. This is a growing proportion of our intake, particularly when the colour chemists, fuel scientists, and environmental scientists are included. We have also found that students even with apparently impressive A-level grades have benefited from this alternative package.

This is not a 'mathematics textbook' and does not try to be. It has a different aim: that of building the confidence of our students so that they are not 'put off' by the mathematics. We have wanted our students to be sufficiently confident that they can see through to the more important chemical principles and ideas behind the equations. Once the fear has been overcome, the necessary skills are built slowly by practice.

The Workbook is intended to be written in. At Leeds, it is used as part of an 18 hour course delivered over 6 sessions in the first 3 weeks of the academic year. The students work in groups, with a Workbook each and so combine a mixture of self-pacing with staff and peer support. Tutors can monitor progress as the booklet is completed. It is also important that this is not a 'stand alone' book: it is envisaged that students may wish to work through the material alongside their chemistry lectures, so that they can see direct applications and get support as and when it is needed.

It is a pleasure to acknowledge the help of the various Chemistry undergraduates who have helped the course improve as they have gone through it, and Drs Gwen Pilling, Barry Johnson, Chris Hildyard, Joel Smith and Mrs. Elin Carson for their comments at various stages.

Leeds   S. K. Scott
June 1995

Contents

SECTION 1

Warm-Up Exercises: Algebra – Substitution and Rearrangement

Algebra – Substitution and Rearrangement

1.1 Getting started: adding and multiplying

Despite their familiarity, it will help if we begin by reviewing the most basic rules of addition and multiplication. The following points will form the basis of almost all the material in this workbook.

> The order in which quantities are added together is unimportant

thus $\qquad$ $3 + 7 = 10 \qquad$ and $\qquad 7 + 3 = 10$

or, more generally, $\qquad x + y = y + x$

Similarly, the order in which numbers are subtracted is unimportant.
Provided we are careful to avoid a few traps, this rule also applies to multiplication and division.

> Numbers can be added together and the sum multiplied by another quantity or they can each be multiplied by the quantity first and then added together, giving the same result.

i.e. $\qquad$
$$10 \times (3 + 7) = 10 \times (10) = 100$$
$$10 \times 3 + 10 \times 7 = 30 + 70 = 100$$

or, more generally, $\qquad a \times x + a \times y = a \times (x + y)$

In this case, we say that there is a *common factor a* in the equation.

Factorisation and *multiplying out* terms in an equation make use of these rules.

Example

We can factorise the equation $y = 9x + 6$ by noting that both 9 and 6 are divisible by 3, to give $y = 3(3x + 2)$. Here the 'common factor' is 3.

This sort of factorisation is useful in tidying-up equations, but not always vital: the first form of the equation is still completely correct as an answer.

Example

We can multiply the equation $y = 5(2x + 3)$ out to give $y = 10x + 15$.

This multiplying out process is valuable when more than one term involves the quantity x.

Example

we can multiply out the equation

$$y = (5 + x)(x + 3).$$

First, we can re-write the right-hand side as

$$5 \times (x + 3) \ + \ x \times (x + 3)$$

so that we have taken each of the two parts of the first bracket separately.
   Next, we multiply out each of the two terms to give

$$(5 \times x + 5 \times 3) + (x \times x + x \times 3), \quad \text{i.e. } 5x + 15 + x^2 + 3x.$$

There are two terms here that involve x multiplied by a number, so we can finish by adding these and re-ordering the terms (for tidiness) to give

$$y \ = \ x^2 \ + \ 8x \ + \ 15.$$

A more general statement of this process is

$$(x + a)(x + b) \ = \ x^2 \ + \ (a + b)x \ + ab$$

(The notation ab is frequently used as short-hand for $a \times b$, just as ax means $a \times x$.)

The previous example had $a = 5$ and $b = 3$, so $a + b = 8$ and $ab = 15$.

The reverse process, factorising an equation into terms of the form $(a + x)$ multiplied together is a valuable art that comes mainly with practice. If the same factors can be found in different parts of the equation, they can frequently be cancelled to bring about simplifications.

Exercises

Multiply out the following equations: in (a) use the method demonstrated above in full; for (b) and (c) use the general result directly. There is a minor adaptation needed in (d). Exercise (e) has three terms and so will lead to a cubic equation: it is best to do this in two stages, taking the first two terms and then multiplying out this result with the third term.

(a) $y = (x + 1)(x + 2)$

(b) $y = (x + 9)(x + 11)$

(c) $y = (x + 1)(x - 2)$

(d) $y = (3x + 1)(x + 2)$

(e) $y = (x + 1)(x + 2)(x + 3)$

(Answers: (a) $x^2 + 3x + 2$, (b) $x^2 + 20x + 99$, (c) $x^2 - x - 2$, (d) $3x^2 + 7x + 2$, (e) $x^3 + 6x^2 + 11x + 6$)

Two special cases are worth remembering.

The first arises when we wish to obtain the square of a term such as $(x + a)$. In that case we want $(x + a) \times (x + a)$, which is the above problem with $b = a$. The result for the square can then be found

$$(x + a)^2 = x^2 + 2ax + a^2$$

The second occurs if $b = -a$, i.e. when $(x + a)$ is multiplied by $(x - a)$

$$(x + a)(x - a) = x^2 - a^2$$

Exercise

Expand the following by multiplying out

(a) $(x + 1)^2 =$

(b) $(x + 3)^2 =$

(c) $(x - 1)^2 =$

(d) $(x + 4)(x - 4) =$

(Answers: (a) $x^2 + 2x + 1$, (b) $x^2 + 6x + 9$, (c) $x^2 - 2x + 1$, (d) $x^2 - 16$.)

We can repeat this method to include situations where more terms are multiplied together

Exercise

Expand the cubic form

$$(x + 1)^3 =$$

(Answer: $x^3 + 3x^2 + 3x + 1$)

In general

$$(x + a)^3 = x^3 + 3ax^2 + 3a^2x + a^3$$

1.2 Adding and subtracting negative numbers

The basic rules are

Adding a negative number or quantity is like subtracting a positive number
Subtracting a negative number or quantity is like adding a positive number

---

Example

If $x = 7$ and $y = -3$, then
$$x + y = 7 - 3 = 4, \quad x - y = 7 - (-3) = 7 + 3 = 10$$

This rule is frequently needed in chemical thermodynamics when calculating overall enthalpy or free energy changes. For a chemical reaction the rule for determining the overall enthalpy change $\Delta_{rxn}H$ is often written in the form

$$\Delta_{rxn}H = \sum_{products}\Delta_f H - \sum_{reactants}\Delta_f H$$

Here the $\Delta_f H$ are the enthalpies of formations of the chemical species involved in the reaction and the symbols Σ (summation) are instructions to add together the enthalpies of formation for the products and the reactants respectively. As enthalpies of formation can be either positive or negative, we frequently have to add and subtract negative numbers in this process.

Examples

(i) calculate the overall enthalpy change for the reaction

$$HCl(g) + Cl(g) \rightarrow H(g) + Cl_2(g)$$

using the following information

	H(g)	Cl_2(g)	HCl(g)	Cl(g)
$\Delta_f H$/kJ mol^{-1}	218	0	-92	122

H(g) and Cl_2(g) are the products, whilst HCl(g) and Cl(g) are the reactants here, so the equation for $\Delta_{rxn}H$ can now be written as

$$\Delta_{rxn}H = (218 + 0) - (122 - 92) = 218 - 30 = 188\,\text{kJ mol}^{-1}$$

The reaction is endothermic as $\Delta H > 0$.

(ii) calculate the overall enthalpy change for the reaction

$$HCl(g) + F(g) \rightarrow HF(g) + Cl(g)$$

using the following information

	HF(g)	Cl(g)	HCl(g)	F(g)
$\Delta_f H$/kJ mol^{-1}	-271	122	-92	79

Proceeding as before,

$$\Delta_{rxn}H = (-271 + 122) - (-92 + 79) = -149 - (-13) = -149 + 13 = -136\,\text{kJ mol}^{-1}$$

In this case the reaction is exothermic as $\Delta_{rxn}H < 0$.

Exercises

Use the enthalpies of formation provided to calculate the overall enthalpy change in each of the following reactions:

(a) $Cl(g) + HF(g) \rightarrow HCl(g) + F(g)$

	HCl(g)	F(g)	HF(g)	Cl(g)
$\Delta_f H$/kJ mol^{-1}	−92	79	−271	122

comment on this result compared to example (ii) above.

(b) $HCl(aq) + NaOH(s) \rightarrow NaCl(s) + H_2O(l)$

	NaCl(s)	H$_2$O(l)	HCl(aq)	NaOH(s)
$\Delta_f H$/kJ mol^{-1}	−411	−286	−167	−426

(c) $CH_4(g) + 2O_2(g) \rightarrow CO_2(g) + 2H_2O(l)$

	CO$_2$(g)	H$_2$O(l)	CH$_4$(g)	O$_2$(g)
$\Delta_f H$/kJ mol^{-1}	−394	−286	−75	0

this is the enthalpy of combustion of methane: the stoichiometric equation shows that two molecules of O_2 are consumed and two molecules of water produced.

(d) $2HCN(g) + 2.5\,O_2(g) \rightarrow 2CO_2(g) + H_2O(l) + N_2(g)$

	CO$_2$(g)	H$_2$O(l)	N$_2$(g)	HCN(g)	O$_2$(g)
$\Delta_f H$/kJ mol^{-1}	−394	−286	0	135	0

(Answers: (a) +136 kJ mol^{-1}, reversing the direction of a reaction alters the sign of the enthalpy change, (b) −104 kJ mol^{-1}, (c) −891 kJ mol^{-1}, (d) −1344 kJ mol^{-1})

1.3 Adding and multiplying fractions

There is a common mistake that leads to much grief when fractions have to be added or subtracted that can be easily sorted out at this stage.

If we have two fractions $1/a$ and $1/b$ that we wish to add, the following is **not true**

$$\frac{1}{a} + \frac{1}{b} = \frac{1}{a+b} \qquad \textbf{(wrong, wrong, wrong!)}$$

We can only add fractions if they have the same denominator (the part underneath the line).

If a and b are already the same, we can add them, in which case we would get $2/a$ or $2/b$ (which are the same if $a = b$).

In general, however, a and b will be different. But we can make them the same by a rule that we will work out in full in the next section:

a fraction is not altered if the top and bottom are multiplied by the same quantity:

$$\text{thus} \qquad \frac{1}{a} = \frac{x}{ax} \qquad \text{for any value of } x$$

(mathematical pedants will point out that this does not work if $x = 0$, division by zero is always a difficult case but not be of concern here).

The way we use this is to multiply the top and bottom of $1/a$ by b, to give b/ab and to multiply the top and bottom of $1/b$ by a to give a/ab. These two terms then have the same group ab in the denominator and so can be added

$$\frac{1}{a} + \frac{1}{b} = \frac{b}{ab} + \frac{a}{ab} = \frac{b+a}{ab} \quad \textbf{(right!)}$$

The same rule applies when we are subtracting fractions, so

$$\frac{1}{a} - \frac{1}{b} = \frac{b}{ab} - \frac{a}{ab} = \frac{b-a}{ab}$$

Exercises

Evaluate the following
   (a) $1/5 + 1/6 =$

   (b) $1/5 + 1/x =$

   (c) $1/5 + 2/x =$

   (d) $2/x + 3/y =$

   (e) $1/5 - 1/6 =$

(f) $2/x - 1/y =$

(g) $1/(x+1) + 1/(x+2) =$

(Answers: (a), 11/30; (b), $(5 + x)/5x$; (c), $(10 + x)/5x$; (d), $(2y + 3x)/xy)$; (e), 1/30; (f), $(2y - x)/xy$; (g), $(2x + 3)/(x+1)(x+2)$.)

Multiplying fractions is relatively straightforward and proceeds by the rule

$$\frac{a}{x} \times \frac{b}{y} = \frac{ab}{xy}$$

Dividing by a fraction is the same as multiplying by the inverse of the fraction, thus

$$x \div \frac{a}{b} = x \times \frac{b}{a} = \frac{bx}{a} \qquad \text{and} \qquad \frac{a}{x} \div \frac{b}{y} = \frac{a}{x} \times \frac{y}{b} = \frac{ay}{bx}$$

Note: the second of these equations can also be written as

$$\frac{a/x}{b/y} = \frac{ay}{bx}$$

for example

$$\frac{1/2}{3/5} = \frac{1 \times 5}{2 \times 3} = \frac{5}{6}$$

Also, for the special case $y = 1$

$$\frac{a/x}{b} = \frac{a}{bx}$$

Exercises

Complete the following

(a) $\quad \frac{2}{3} \times \frac{5}{7} =$

(b) $\quad \frac{2}{3} \times \frac{5}{8} =$

(c) $\quad \frac{(1+x)}{(2+x)} \times \frac{(1+x)}{(4+x)} =$

(d) $\quad \frac{2}{3} \div \frac{5}{7} =$

(e) $\quad \frac{(1+x)}{(2+x)} \div \frac{(1+x)}{(4+x)} =$

(Answers: (a) $\frac{2\times5}{3\times7} = \frac{10}{21}$; (b) $\frac{2\times5}{3\times8} = \frac{10}{24} = \frac{5}{12}$, cancelling the common factor in the numerator and denominator; (c) $(1+x)^2 / (2+x)(4+x)$; (d) $\frac{2}{3} \div \frac{5}{7} = \frac{2}{3} \times \frac{7}{5} = \frac{14}{15}$;

(e) $\frac{(1+x)}{(2+x)} \div \frac{(1+x)}{(4+x)} = \frac{(1+x)}{(2+x)} \times \frac{(4+x)}{(1+x)} = \frac{(1+x)(4+x)}{(2+x)(1+x)} = \frac{(4+x)}{(2+x)}$

with the factor (1+x) cancelling from the numerator and denominator.)

1.4 General rules of algebra

There are two simple but important rules covering the basic features of algebra:

> An equation is unaffected if the same quantity is added or subtracted from each side
> An equation is unaffected if each side is multiplied or divided by the same quantity

These are used, with the rules governing how different terms in an equation cancel each other out, to simplify and re-arrange equations

> The same quantity can be added to each side of an equation

for example: if $x = y$,   then $x + 3 = y + 3$

We use this with the rule:

> If the same quantity arises twice on the same side of an equation, once with a + sign and once with a − sign, then these will cancel.

For example, if we know that $x + 7 = y$, then we can use these rules to obtain an expression for x. We want to 'move' the 7 over to the 'other side' of the equation. We can do this by *subtracting* 7 from each side:

$$x + 7 - 7 = y - 7$$

so then cancelling the +7 and the −7 on the left-hand side we have

$$x = y - 7$$

The quantity that appeared with a + sign on one side has moved to the other with a − sign.

> Both sides of an equation can be multiplied by the same quantity

for example   if $x = y$,   then $3x = 3y$

We use this with the rule:

> Any term occurring as a factor on both the top and the bottom of the same side of an equation can be cancelled.

For example, if we have the equation $7x = y$, we can use this approach to determine x in terms of y: we divide both sides by 7, to give

$$\frac{7x}{7} = \frac{y}{7}$$

then, cancelling the 7's in the top (numerator) and bottom (denominator) of the left-hand side, we obtain

$$x = \tfrac{y}{7}$$

The factor of 7 has gone from the numerator on one side (multiplying x) to the denominator of the other (dividing y).

These two examples are rather trivial, but they do form the basis of much that the chemist is required to use.

We can see the two rules working together in the following

Example

The definition of the change in the Gibbs function ΔG in terms of the enthalpy change ΔH, entropy change ΔS and temperature T is

$$\Delta G = \Delta H - T\Delta S$$

We can rearrange this to find an expression for ΔS:

stage 1: move the terms $T\Delta S$ and ΔG to the opposite sides by *subtracting* ΔG from each side and *adding* the term $T\Delta S$ to each side:

$$\Delta G - \Delta G + T\Delta S = \Delta H - T\Delta S - \Delta G + T\Delta S$$

stage 2: cancel terms on each side — the ΔG and $-\Delta G$ cancel on the left-hand side; the $-T\Delta S$ and $+T\Delta S$ cancel on the right-hand side, to give

$$T\Delta S = \Delta H - \Delta G$$

stage 3: divide both sides by T and then cancel this from the top and bottom of the left-hand side to give

$$\Delta S = (\Delta H - \Delta G)/T$$

Exercises

(a) the velocity c, frequency ν and wavelength λ of a wave are related by $c = \nu\lambda$. Re-arrange this to find the expressions for (i) ν, and (ii) λ

(i) $\nu =$

(ii) $\lambda =$

(b) The *wavenumber* $\bar{\nu}$ is defined as $\bar{\nu} = 1/\lambda$ Express $\bar{\nu}$ in terms of c and ν $\bar{\nu} = \dfrac{1}{\lambda} =$

(c) the energy ε of a photon is given by $\varepsilon = h\nu$, where ν is the frequency and h is Planck's constant. Express ε in terms of (i) the wavelength and (ii) the wavenumber.

(i)

(ii)

(d) The ideal gas equation can be written in the form $pV = nR(\theta + 273.15)$, where θ represents the temperature on the Celsius scale. Rearrange this to given an equation for θ.

(Answers: (a) (i) $\nu = c/\lambda$, (ii) $\lambda = c/\nu$, (b) $\bar{\nu} = \nu/c$, (c) (i) $\varepsilon = hc/\lambda$, (ii) $\varepsilon = hc\,\bar{\nu}$; (d) $\theta = (pV/nR) - 273.15$.)

There are other processes that can be applied to equations

Both sides of an equation can be **inverted**, i.e. we can take the reciprocal each side and the resulting terms will also be equal

For instance, if $x = 10$, then $1/x = 1/10$.

Again, there are some common mistakes made inverting equations when there is more than one term on one side of the equation:

If $\qquad x = a + b \qquad$ then $\qquad \dfrac{1}{x} = \dfrac{1}{a+b} \qquad$ **not** $\qquad \dfrac{1}{a} + \dfrac{1}{b}$

Similarly, if $\qquad \dfrac{1}{x} = \dfrac{1}{a} + \dfrac{1}{b}$

then $\qquad x = \dfrac{1}{\dfrac{1}{a} + \dfrac{1}{b}} = \dfrac{ab}{a+b} \qquad$ **not** $\qquad a + b$

Example

An example of this arises in the kinetics of second-order reactions for which the concentration c of some species varies in time according to the equation

$$\frac{1}{c} = \frac{1}{c_0} + kt$$

where c_0 is the initial concentration and k is the reaction rate constant.

First, we invert each side to give

$$c = \frac{1}{\frac{1}{c_0} + kt}$$

We have to take one over **the whole** of the right-hand side. To simplify the denominator further, we can multiply the top and bottom of the right-hand side by c_0 to give

$$c = \frac{c_0}{c_0\left(\frac{1}{c_0} + kt\right)} = \frac{c_0}{\left(\frac{c_0}{c_0} + c_0 kt\right)} = \frac{c_0}{1 + c_0 kt}$$

Exercises

(a) The ideal gas equation has the form $pV = nRT$. Rearrange this to find the appropriate expression for the molar volume $V_m = V/n$

$$V/n =$$

(b) The van der Waals equation of state has the form $\left(p + \dfrac{a}{V_m^2}\right)(V_m - b) = RT$.

Find the expression for p

$$p =$$

(c) The spectroscopic lines in the Balmer series for the hydrogen atom occur at wavenumbers given by the following equation

$$\bar{v} = \Re_H\left(\frac{1}{4} - \frac{1}{n^2}\right)$$

where $\Re_H$ is the Rydberg constant and n is the quantum number of the state from which the transition is occurring.

(i) Rearrange this to give an equation for $1/n^2$

(ii) Invert this to obtain the expression for n^2.

(Answers: (a) $V_m = RT/p$, note that the inverse of this quantity, $n/V = p/RT$ is the *concentration*;

(b) $p = \dfrac{RT}{V_m - b} - \dfrac{a}{V_m^2}$; (c) (i) $\dfrac{1}{n^2} = \dfrac{1}{4} - \dfrac{\bar{v}}{\Re_H} = \dfrac{\Re_H - 4\bar{v}}{4\Re_H}$, (ii) $n^2 = 4\Re_H / (\Re_H - 4\bar{v})$.)

1.5 Substituting into equations: simultaneous equations

Often in chemistry we have two equations involving the same quantity. It is then possible to substitute from one equation into the other to eliminate that quantity.

Example

The mean kinetic energy KE of a molecule of mass m is given by

$$KE = \tfrac{1}{2}mc^2$$

where c is the root mean square velocity.

From the equipartition theorem, the mean kinetic energy of a monoatomic gas is also given by

$$KE = \tfrac{3}{2}kT$$

where k is the Boltzmann constant and T the temperature.

As KE is the same in these equations, the two right-hand sides must also be equal. Thus we can substitute to give

$$\tfrac{1}{2}mc^2 = \tfrac{3}{2}kT \qquad \text{so} \qquad c = \sqrt{3kT/m}$$

Exercises

(a) The Gibbs function G is given by $G = H - TS$. The enthalpy H is given by $H = U + pV$.

Eliminate H from the first equation

$$G =$$

(b) The equilibrium constant K for a reaction is given by the equation $\ln K_{eq} = -\dfrac{\Delta G^{\theta}}{RT}$

Use the equation $\Delta G^{\theta} = \Delta H^{\theta} - T\Delta S^{\theta}$ to eliminate ΔG^{θ}.

$$\ln K_{eq} =$$

(Answers: (a) $G = U + pV - TS$; $\ln K_{eq} = -\dfrac{\Delta H^{\theta}}{RT} + \dfrac{\Delta S^{\theta}}{R}$.)

This method also forms the basis of the route to solving *simultaneous equations*. These arise when we have two different equations relating two unknown quantities x and y. We can use these two equations to solve for the two unknowns. (In general we need n equations to find n unknown quantities.)

If we have two equations of the form

$$y = ax + b$$

and

$$y = cx + d$$

where a, b, c and d are known numbers, we can find y and x as follows:

As both right-hand sides are equal to the same thing (i.e. $= y$), they must also be equal to each other, so we can write

$$ax + b = cx + d$$

Subtracting b from each side and then subtracting cx from each side, we then have

$$(a - c)x = (d - b)$$

so

$$x = \frac{d - b}{a - c}$$

We can then find y by substituting this result for x into either original equation

$$y = a\left(\frac{d - b}{a - c}\right) + b = \left(\frac{ad - bc}{a - c}\right)$$

Check that you can do the simplification giving the second form in this equation. (Multiply and divide the final b by $(a - c)$ so that the two terms on the right-hand side then have the same denominator and so can be added. Multiplying out the terms in the numerator and cancelling where possible leads to the second form.)

Exercise

Find the values for x and y in the following pairs of equations

(a) $y = 3x + 5$

         $y = 2x + 9$

(b) $y = 3x + 5$

         $y = 2x - 9$

(c) $3y = 9x + 15$

         $y = 2x + 9$

(Answers: (a) using $a = 3$, $b = 5$, $c = 2$, $d = 9$, we have $x = (9 - 5)/(3 - 2) = 4$ and $y = (3 \times 9 - 2 \times 5)/(3 - 2) = 17$; (b) $x = -14$, $y = -37$, (c) divide the first equation through by 3, so we just have y on the right-hand side — the equations are then the same as in (c) and so has the same solution.)

In practice, the main trick with such simultaneous equations lies in identifying which terms are known, and hence correspond to a, b, c and d, and which are the 'unknowns' x and y.

Exercise

The Arrhenius equation for a reaction rate constant k can be written in the form

$$\ln k = \ln A - \frac{E}{R}\frac{1}{T}$$

where T is the temperature, E is the activation energy and A the pre-exponential factor; R is the gas constant.

(i) At how many different temperatures must measurements of k be made in order to evaluate A and E?

(ii) The following data were obtained

$$\ln k = 14.0 \text{ at } 300 \text{ K}$$
$$\ln k = 14.7 \text{ at } 320 \text{ K}$$

Substitute the values for $\ln k$ and $1/T$ into the Arrhenius equation for each case:

$$= \ln A - \qquad \frac{E}{R}$$

$$= \ln A - \qquad \frac{E}{R}$$

(iii) Now identifying $\ln A$ as y and E/R as x (the unknowns), rearrange these equations to the form $y = ax + b$ and $y = cx + d$.

(iv) Solve for $\ln A$ and E/R

(Answers: (i) there are two unknowns ($\ln A$ and E/R) so we need two equations and, hence, two measurements of k at different temperatures, (ii) $14.0 = \ln A - \frac{1}{300}\frac{E}{R}$, $14.7 = \ln A - \frac{1}{320}\frac{E}{R}$;

(iii) $y = \frac{1}{300}x + 14.0$, $y = \frac{1}{320}x + 14.7$ so a = 1/300, b = 14.0, c = 1/320 and d = 14.7,

(iv) $x = 3360$, so $E = 27.9$ kJ mol^{-1}, and $y = \ln A = 25.2$.)

Exercise

The variation of the reaction equilibrium constant with temperature can be written in the form

$$\ln K_{eq} = -\frac{\Delta H^{\theta}}{R}\frac{1}{T} + \Delta S^{\theta}$$

Use the method developed above to determine the enthalpy and entropy changes from the following two measurements

$$\ln K_{eq} = 4 \text{ at } T = 500 \text{ K}$$

$$\ln K_{eq} = -4 \text{ at } T = 1000 \text{ K}$$

(Answers: $\Delta H^{\theta}/R = -8000$, $\Delta S^{\theta} = -12$.)

Exercise

The absorbance A of a solution containing an unknown mixture of Co^{2+} and Ni^{2+} ions is measured in a 1 cm cell at wavelengths $\lambda_1 = 510$ nm and $\lambda_2 = 660$ nm. Use the following information to determine the concentrations of the two ions.

The Beer-Lambert law for a mixture can be written in the following form for each wavelength:

$$A_{\lambda=510nm} = \varepsilon_{\lambda=510nm,Co^{2+}} \ell \left[Co^{2+}\right] + \varepsilon_{\lambda=510nm,Ni^{2+}} \ell \left[Ni^{2+}\right]$$

$$A_{\lambda=660nm} = \varepsilon_{\lambda=660nm,Co^{2+}} \ell \left[Co^{2+}\right] + \varepsilon_{\lambda=660nm,Ni^{2+}} \ell \left[Ni^{2+}\right]$$

where $\varepsilon_{\lambda=510nm,Co^{2+}}$ etc. are the *molar decadic extinction coefficients* for the two ions at the specified wavelengths and ℓ is the path length.

The required data are:

$A_{\lambda=510nm} = 0.611$; $\varepsilon_{\lambda=510nm,Co^{2+}} = 4.85\,mol^{-1}\,dm^3\,cm^{-1}$; $\varepsilon_{\lambda=510nm,Ni^{2+}} = 0.063\,mol^{-1}\,dm^3\,cm^{-1}$;

$A_{\lambda=660nm} = 0.167$; $\varepsilon_{\lambda=660nm,Co^{2+}} = 0.23\,mol^{-1}\,dm^3\,cm^{-1}$; $\varepsilon_{\lambda=660nm,Ni^{2+}} = 1.84\,mol^{-1}\,dm^3\,cm^{-1}$

with $\ell = 1$ cm. (Note that the extinction coefficients have units that mean we will calculate the concentration in mol dm^{-3}).

(a) Substitute the various data values into the Beer-Lambert equation to obtain the simultaneous equations involving [Co^{2+}] and [Ni^{2+}] to be solved

(b) If we decide to identify x = [Co^{2+}] and y = [Ni^{2+}], we should rearrange the equations so they have the form $y = ax + b$ and $y = cx + d$

$y =$

$y =$

(c) now solve to find the unknown concentrations

[Co^{2+}] $= x =$

[Ni^{2+}] $= y =$

(Answers: (a) 0.611 = 4.85[Co^{2+}] + 0.063[Ni^{2+}], 0.167 = 0.23[Co^{2+}] + 1.84[Ni^{2+}]; (b) dividing through by 0.063 in the first equation and by 1.84 in the second and then moving the term in [Co^{2+}] to the other side in each case we get $y = -77.0x + 9.70$ and $y = -0.125x + 0.0908$; (c) [Co^{2+}] = 0.125 mol dm^{-3}, [Ni^{2+}] = 0.075 mol dm^{-3}.)

1.6 Solving quadratic equations

A quadratic equation has the general form

$$ax^2 + bx + c = 0$$

where a, b and c are known numbers or *coefficients*.

Sometimes, it is possible to see that the equation can be factorised into the product of two terms involving x in the form

$$a(x-d)(x-e) = 0$$

where the new coefficients d and e are related to a, b and c.

The equation is *satisfied* when either $x = d$, so the first term is zero, or when $x = e$, so that the second term is zero. Thus, d and e are called the *roots* of the equation, i.e. the values of x for which the equation is true.

Factorisation is a skill that comes with practice and is not always possible simply by looking at the equation. There is, however, a method that will work for all cases

In general, the roots of a quadratic are given by the formula

$$x = \frac{-b \pm \sqrt{b^2 - 4ac}}{2a}$$

This equation is worth memorising.

There are two roots of the quadratic equation, one corresponding to the + sign and one to the − sign in front of the square root term (which is known as the *discriminant*).

Exercise

Find the roots of the following quadratic equations

(i) $x^2 + x - 6 = 0$

(ii) $3x^2 - 13x + 4 = 0$

(Answer: (i) $a = 1$, $b = 1$, $c = -6$, $x = \frac{1}{2}\{-1 \pm \sqrt{25}\} = \frac{1}{2}\{-1 \pm 5\} = 2$ and -3;

(ii) $x = \frac{1}{6}\{13 \pm \sqrt{169 - 48}\} = \frac{1}{6}\{13 \pm 11\} = 4$ or $\frac{1}{3}$.)

Once the roots of an equation have been found, we can write down the factorised form.

Example

In exercise (i) above, the roots were found to be:

$$x = 2 \qquad \text{i.e.} \qquad x - 2 = 0$$

and $\qquad\qquad x = -3 \qquad \text{i.e.} \qquad x + 3 = 0$

so we can write the original equation in the form

$$x^2 + x - 6 = (x-2)(x+3) = 0$$

using the second form of the equation above.

Exercise:

Factorise the equation whose roots were obtained in exercise (ii) above

$$3x^2 - 13x + 4 =$$

(Answer: $\left(x - 4\right)\left(x - \frac{1}{3}\right) = 0$.)

Quadratic equations frequently arise in calculations involving equilibria in chemistry. If we consider the general reaction

$$A + B \rightleftharpoons AB$$

We can denote the initial concentration of A and B as a_0 and b_0, and the equilibrium concentrations of A, B and AB as a, b and x respectively. Because of the reaction stoichiometry, these various concentrations are related, with

$$a = a_0 - x \qquad \text{and} \qquad b = b_0 - x$$

These simply state that the amount of AB formed is equal to the amounts of A and B consumed. We can thus represent the concentrations at equilibrium as

$$A \quad + \quad B \quad \rightleftharpoons \quad AB$$
$$(a_0 - x) \quad (b_0 - x) \qquad\qquad x$$

The equilibrium constant K_{eq} is related to these concentrations by

$$K_{eq} = \frac{[AB]}{[A][B]}$$

so
$$K_{eq} = \frac{x}{(a_0 - x)(b_0 - x)}$$

Exercise

Rearrange the equation for K_{eq} to give a quadratic equation for x

(Hint: multiply the terms in the denominator of the right-hand side up onto the left-hand side, then multiply the brackets out and finally bring the remaining x term over from the right-hand side.)

(Answer: $K_{eq}x^2 - \{1 + K_{eq}(a_0 + b_0)\}x + a_0b_0K_{eq} = 0$.)

If K_{eq} and the initial concentrations are known, we can then find the equilibrium concentration x of AB

Exercise

Determine the equilibrium concentration of A, B and AB if $a_0 = 1$, $b_0 = 1$ and $K_{eq} = 5$ (omitting the units to make the algebra simpler)

Write the quadratic equation with the values of a_0, b_0 and K_{eq} substituted in:

Use the equation for the roots of a quadratic to find x

$$x = \qquad \text{or} \qquad x =$$

One of these roots corresponds to a concentration greater than the initial concentrations of A and B. This is usually the case, that one of the roots is not physically acceptable, and we must choose the other

$$x =$$

Now calculate a and b from x, a_0 and b_0

$$a =$$

$$b =$$

(Answers: $5x^2 - 11x + 5 = 0$, $x = \frac{1}{10}\{11 \pm \sqrt{101}\} = 0.64$ or 1.56; take $x = 0.64$ giving $a = b = 1 - 0.64 = 0.36$.)

Exercise

Repeat the above analysis to find the equilibrium concentration of AB under the same conditions for the reaction

$$A_2 + B_2 = 2\,AB$$

(Partial Answer: This is a real test of your understanding. The first changes arise from the different stoichiometry. As two AB molecules are produced, the concentration will be $2x$ with $a = a_0 - x$ and $b = b_0 - x$. Also the equilibrium constant will have $(2x)^2$ in the numerator $K_{eq} = (2x)^2/(a_0 - x)(b_0 - x)$ rather than just x as in the above example. The quadratic obtained on rearrangement will have a different form

$$\left(K_{eq} - 4\right)x^2 - K_{eq}\left(a_0 + b_0\right)x + K_{eq}a_0 b_0 = 0$$

Taking $K_{eq} = 5$, and $a_0 = b_0 = 1$ as before, we find $x = 0.528$ (or 9.47 which can be discarded). Thus the concentration of A_2, B_2 and AB at equilibrium are 0.472, 0.472 and 1.056 (= $2x$).)

In some cases, the coefficients a, b and c in the quadratic equation are such that the discriminant is negative.

This arises, for example in the equation

$$x^2 + x + 1 = 0$$

for which the roots are given by $x = \frac{1}{2}\left\{-1 \pm \sqrt{-3}\right\}$.

Negative numbers do not have real square roots and the calculation will fail on a standard calculator. The square roots of negative numbers are called *imaginary* and do have some relevance in chemistry: we will deal with these in Section 5.7 and 5.8.

1.7 Proportions and percentages

In the last example, the quantity x was found to be 0.528. This represents the concentration of a product formed from reactants with initial concentrations equal to 1. In the reaction, 1 mole of reactant A_2 forms one mole of product AB if the reaction goes to completion. We can express the actual yield, which here is 0.528 mol, as a *percentage* of the maximum possible yield. In this case, the percentage yield is 52.8%.

In general, if we have two numbers x and y, then we calculate the percentage that x is of y using the formula

$$\text{percentage} = \frac{x}{y} \times 100\%$$

Exercise

What percentage is 37 of 132?

(Answer: $\frac{37}{132} \times 100\% = 28\%$)

The %-sign should always be written out in the answer as this reminds us that we have multiplied the simple fraction by 100%.

To find the number x that is a certain percentage of y, we work in reverse:

Example

What is 31% of 142?

We use $\qquad x = \frac{31\%}{100\%} y = \frac{31\%}{100\%} \times 142 = 0.31 \times 142 = 44.02$

so 44.02 is 31% of 142.

In general

$$x = \frac{\text{percentage}}{100\%} \times y$$

Percentages are important when dealing with errors that arise in experimental measurements and in reporting yields in preparative chemistry.

Example

In a calorimetry experiment, we may measure a temperature rise ΔT of 5.0 K. There will be some uncertainty in this measurement, arising perhaps from the accuracy with which we can read off from the thermometer. This limit of precision might be 0.1 K. To indicate this uncertainty, the result would then be reported as

$$\Delta T = 5.0\left(\pm 0.1\right)K$$

What percentage of the measurement is the uncertainty or 'error'?

$$\text{percentage error} = \qquad \%$$

(Answer: 2%.)

If this temperature rise is now used to calculate the change in internal energy, ΔU, there will be a corresponding uncertainty or error in the result.

Provided ΔT is only multiplied or divided by quantities which are known accurately, then the resulting error will be the same percentage of ΔU as the experimental uncertainty is in ΔT. In other words the error remains the same proportion of both quantities.

So, if the final result is $\Delta U = 100$ kJ mol^{-1}, the error will be $\pm$ 2% of this, i.e. $\pm$ 2 kJ mol^{-1}

$$\Delta U = 100\left(\pm 2\right)kJ\ mol^{-1}$$

Example

The rate constant for a reaction is measured at several temperatures and an Arrhenius plot of ln k versus 1/T is made. The slope is related to the activation energy E by $m = -E/R$ where $R = 8.314$ J K^{-1} mol^{-1} is the gas constant (which is known precisely). The slope m of this graph is measured to be -14400 K (note that this has units of temperature), calculate the activation energy.

The equation we need is $m = -\dfrac{E}{R} = -14400\ K$

so $\qquad\qquad\qquad E\ =\ R \times 14400\ K\ =\ 120\ kJ\ mol^{-1}$

If the uncertainty in the slope had been estimated to be $\pm$ 1000 what would be the corresponding uncertainty in E?

We can find this using the percentage method:

± 1000 in 14400 is a percentage error of $\pm 7\%$.

So there will be an uncertainty of $\pm 7\%$ in the activation energy, i.e. an uncertainty of $\pm$ 8.4 kJ mol^{-1}:

$$E\ =\ 120\ (\pm 8.4)\ kJ\ mol^{-1}$$

In calculating yields in a preparative experiment we need to know both the actual yield and the maximum possible yield in any particular reaction.

Example

In the esterification reaction

$$CH_3OH + CH_3COOH \rightarrow CH_3OOCCH_3 + H_2O$$

the reactants combine in the following proportions, based on their molar masses: 32 g of CH_3OH reacts with 60 g of CH_3COOH to form 74 g of the ester product.

In a particular reaction 1.6 g of methanol is mixed with an excess of ethanoic acid. At the end of the experiment 2.5 g of the ester is extracted. What is the percentage yield?

$$\text{percentage yield} = \frac{\text{actual yield}}{\text{theoretical maximum yield}} \times 100\%$$

The theoretical maximum yield is calculated using the proportion relationships:

32 g of CH_3OH yields 74 g ester

1 g of CH_3OH yields $\dfrac{74}{32}$ g ester   (divide through by 32)

1.6 g of CH_3OH yields $\dfrac{1.6g}{32g} \times 74g = 3.7g$ of ester   (multiply through by 1.6)

so   percentage yield $= \dfrac{2.5g}{3.7g} \times 100\% = 67.6\%$

Exercises

(a) Calculate the percentage yield in the reaction

$$C_2H_5NH_2 + CH_3CO_2H \rightarrow C_2H_5NHCOCH_3 + H_2O$$

if 5 g of CH_3CO_2H in an excess of amine yields 4 g of the amide product.

(b) A plot of $\ln(p)$ versus $1/T$ is made of vapour pressure measurements and the slope of this graph used to evaluate the enthalpy of vaporisation $\Delta_{vap}H$. If the uncertainty in the slope is ±100 in 3000 and the enthalpy change is calculated to be 52 kJ mol^{-1}, what is the uncertainty in $\Delta_{vap}H$?

(Answers: (a) 55%, theoretical maximum yield = (5/60) × 87 g = 7.25 g); (b) uncertainty in slope = (100/3000) × 100% = 3.33%, there is the same percentage uncertainty in the final quantity, so $\Delta_{vap}H = 52$ (±1.7) kJ mol^{-1}.)

The rule of proportions can also be used to calculate how much of one reagent is required to react in stoichiometric proportion with a given amount of another reactant.

Example

If we return to the reaction between methanol and ethanoic acid, then as the stoichiometry is one mole of CH_3OH reacting with one mole of CH_3CO_2H, we know that 32 g of methanol reacts with 60 g of ethanoic acid.

How much methanol is required to react in stoichiometric proportion with 7.5 g of CH_3CO_2H?

To calculate this, we can re-write the above statement in the form

$$60 \text{ g } CH_3CO_2H \quad \text{reacts with} \quad 32 \text{ g } CH_3OH$$

This has a similar form to a mathematical equation except that = is replaced by the word 'reacts with'. We can even treat it like an equation. For instance, we can divide through by 60 g to give

$$1 \text{ g of } CH_3CO_2H \text{ reacts with } \frac{32}{60} \text{ g } CH_3OH$$

We can now multiply both sides by 7.5 to find the mass of methanol required to react with 7.5 g of CH_3CO_2H

$$7.5 \text{ g of } CH_3CO_2H \text{ reacts with } \frac{32}{60} \times 7.5 \text{ g } CH_3OH$$

From which we obtain the result that 4 g of methanol is required.

Exercise

Calculate the mass of CO_2 required to react in stoichiometric proportion with 10 g of CaO in the reaction

$$CaO + CO_2 \rightarrow CaCO3_3$$

(Answer: $M_r(CaO) = 56$, $M_r(CO_2) = 44$, so mass required = $(10/56) \times 44g = 7.86g$.)

Summary of Section

The material in this section has covered the most important rules for 'doing algebra' on equations that arise in chemistry. The following should be familiar:

- factorisation and multiplying out $(x + a) \times (x + b) = x^2 + (a + b)x + ab$

- subtracting negative numbers $x - (-a) = x + a$

- adding, multiplying and dividing fractions $\dfrac{1}{a} + \dfrac{1}{b} = \dfrac{a+b}{ab}$, $\dfrac{a}{x} \times \dfrac{b}{y} = \dfrac{ab}{xy}$; $\dfrac{a/x}{b/y} = \dfrac{ay}{bx}$

- rearranging equations $y = x + a$ so $x = y - a$; $y = ax$ so $x = y/a$

- solving simultaneous equations $y = ax + b \quad y = cx + d$

$$x = \frac{d-b}{a-c} \qquad y = \left(\frac{ad-bc}{a-c} \right)$$

- solving quadratic equations $ax^2 + bx + c = 0 \qquad x = \dfrac{-b \pm \sqrt{b^2 - 4ac}}{2a}$

- proportions and percentages $\text{percentage} = \dfrac{x}{y} \times 100\%$

$$x = \frac{\text{percentage}}{100\%} \times y$$

$$\text{percentage yield} = \frac{\text{actual yield}}{\text{theoretical maximum yield}} \times 100\%$$

Gentle Stretching Exercises: Powers, Exponentials and Logarithms

Powers, Exponentials and Logarithms

2.1 Revision: powers of 10

Rather than write numbers with strings of zeros either before or after the decimal point, it is usual to find large or small numbers written as a 'normal-sized' number multiplied by an appropriate power of ten,

---

Example

the number 300 can be written as 3×10^2

the number of molecules in a mole of substance is 6.022×10^{23}

the mass of a proton is 1.673×10^{-27} kg

In each case, the power of 10 indicates how many times the number should be multiplied by 10:

$$10^1 = 10$$
$$10^2 = 10 \times 10 \ (= 100)$$
$$10^3 = 10 \times 10 \times 10 \ (= 1000)$$

In general, 10^n denotes multiplication n-times by ten.

A negative *exponent*, as in 10^{-3}, indicates division by the appropriate power of 10

$$10^{-1} = \frac{1}{10} = 0.1$$
$$10^{-2} = \frac{1}{10^2} = \frac{1}{100} = 0.01$$
$$10^{-3} = \frac{1}{10^3} = \frac{1}{1000} = 0.001$$

---

so $\qquad 10^{-n} = 1/10^n$.

Exercise

Write out the full form for (i) the number of molecules in one mole of substance and (ii) the mass of a proton:

(i) 6.022×10^{23} =

(ii) 1.673×10^{-27} kg =

(Answers: (i) 602 200 000 000 000 000 000 000; (ii) 0.000 000 000 000 000 000 000 000 001 673 kg.)

Exercises

Express the following numbers in their full form:

(i) 2.998×10^{8}

(ii) 9.6485×10^{4}

(iii) 9.109×10^{-31}

Express the following numbers in terms of the appropriate powers of 10

(iv) 0.000 000 000 000 000 000 000 000 000 000 000 662 6

(v)   101 325

(Answers: (i) 299 800 000; (ii) 96 485; (iii) 0.000 000 000 000 000 000 000 000 000 000 910 9;
(iv) 6.626×10^{-34}; (v) 1.01325×10^{5}.)

Very Important Note

The notation 10^{n} is commonly found in textbooks. This number is actually 1×10^{n}.

When entering such a factor into a calculator, make sure that you enter 1E0n rather than 10E0n (which is actually 10×10^{n} or 10^{n+1})

e.g. 1×10^{3} should be entered as 1E+3
similarly 1×10^{-3} should be entered as 1E−3

 Page 30 — Beginning Mathematics for Chemistry

Exercise

Use your calculator to express the molar mass of sucrose in kilograms using the information M(sucrose) = 342.3 g mol⁻¹ and 1 g = 10⁻³ kg

(Answer: enter 342.3 × 1E−3: M(sucrose) = 0.3423 kg mol⁻¹·)

The SI system of notation emphasises the powers of 10³ and 10⁻³ and multiples of these such as 10⁶, 10⁹, 10⁻⁶, 10⁻⁹ etc.

These various powers of 10 are each denoted by a special prefix and associated symbol; common powers in chemistry are given in the table below:

power	prefix	symbol	power	prefix	symbol
10^1	deca	da	10^{-1}	deci	d
10^2	hecto	h	10^{-2}	centi	c
10^3	kilo	k	10^{-3}	milli	m
10^6	mega	M	10^{-6}	micro	μ
10^9	giga	G	10^{-9}	nano	n
10^{12}	tera	T	10^{-12}	pico	p
			10^{-15}	femto	f

(The prefix and symbol for 10^{-1} and 10^{-2} are also included in the above table as these are relevant and occur relatively frequently in chemistry texts.)

Worked Exercise

The bond length R in the molecule HCl is 0.000 000 000 12745 m.
Express this in terms of the most appropriate prefix symbols.

The value for R can be written as

$$R = 0.12745 \times 10^{-9}\,\text{m} \quad \text{or} \quad 127.45 \times 10^{-12}\,\text{m}$$

so $R = 0.12745$ nm or $R = 127.45$ pm.

There is some flexibility for personal choice here as quoting the result in either nanometres or picometres is equally valid: the latter is perhaps more common in the case of bond lengths although the unit of angstrom ($1\text{Å} = 10^{-10}$ m) is also used — the above result being 1.2745 Å

Exercises

Express the following quantities in an appropriate form based on the above prefixes:

(i) the enthalpy of formation of $H_2O(g)$ at 298 K
$$\Delta_f H^\theta(H_2O, \text{ g, } 298 \text{ K}) = -241\ 820 \text{ J mol}^{-1} =$$

(ii) the enthalpy of combustion of propane at 298 K
$$\Delta_{comb} H^\theta(\text{propane, g, } 298 \text{ K}) = -2\ 220\ 000 \text{ J mol}^{-1} =$$

(iii) the wavelength of blue light
$$\lambda = 0.000\ 000\ 47 \text{ m} =$$

(Answers: (i) −241.82 kJ mol^{-1}; (ii) −2.22 MJ mol^{-1}; 470 nm or 0.47 μm.)

2.2 Exponents

The concept of *the square* of a number is reasonably familiar:

$$y^2 = y \times y$$

The *square root* of a number y is the number z that gives y when it is squared:
$$\text{if } \sqrt{y} = y^{1/2} = z \quad \text{then} \quad z^2 = y$$

Any number can be squared.

All squares are positive

For any positive number there are two square roots: one positive and one equal in magnitude but negative

The square root of a negative number is not real

The 'square' and 'square root' commands on calculators are usually on the same button, with one accessed via the 'second function' key.

The *cube* of a number is similarly given by $y^3 = y \times y \times y$.

The *cube root* z of y is defined by
$$z = \sqrt[3]{y} = y^{1/3} \quad \text{so} \quad y = z \times z \times z$$

Many calculators have a cube and cube root function key.

Cube roots of negative numbers are real and are themselves negative.

The notation above can be generalised to compute the n-th power and n-th root of any number, y:

$$n\text{-th power} \qquad y^n = y \times ... \times y$$

where there are n y's multiplied together on the right-hand side.

$$n\text{-th root} \qquad \sqrt[n]{y} = y^{1/n}$$

If $y^{1/n} = z$ then $y = z \times ... \times z$.

Most calculators have a key, variously denoted x^y or y^x, which can be used to compute n-th powers and (using the second function key $x^{1/y}$ or $y^{1/x}$) n-th roots.

Worked Examples

Calculate the 5-th power and 5-th root of 4.

In the above notation, we are seeking the value of 4^5 and then finding the number z such that $z^5 = 4$.

With a calculator equipped with a function key marked x^y, then $x = 4$ and $y = 5$ in this problem. Entering the following keystrokes:

$$4 \; x^y \; 5 =$$

gives the result 1024, which we can check simply by entering $4 \times 4 \times 4 \times 4 \times 4 =$.

To find the 5-th root with the inverse or second function key marked $x^{1/y}$, we enter:

$$4 \; (\text{2nd function/inverse}) \; x^{1/y} \; 5 =$$

to give $z = 1.3195079$.

We can check the root calculation by finding 1.3195079^5 using the above method with the x^y key.

With the newer 'direct algebraic logic' calculators there is a slight difference in the order in which the **numbers** must be entered. For example, to find a root using the $\sqrt[x]{}$ key, we enter

$$5 \; \sqrt[x]{} \; 4 \; \text{Execute}$$

to obtain the fifth root, $4^{1/5}$.

This operation then shows that we can find the n-th power (and the n-th root) of any number, not just integers.

More importantly, the power or *exponent* n also can have non-integer values.

Exercises

Use the method given above to evaluate the following:

(i) $3.2^{1.2} =$

(ii) $\pi^3 =$

(iii) $2.71828^{0.125} =$

(iv) $4.182^{-0.75} =$

(Answers: (i) 4.038; (ii) 31.006; (iii) 1.133; (iv) 0.3419.)

Notes

Exercise (iii) above is actually calculating the 8-th root of the number 2.71828, as $0.125 = \frac{1}{8}$;

A negative value for the exponent is also allowed, as in example (iv); the relationship between y^n and y^{-n} follows that given in the previous section for the special case $y = 10$, i.e.

$$y^{-n} = \frac{1}{y^n}$$

so exercise (iv) can be solved by finding $4.182^{0.75}$ and then taking the inverse (the $1/x$ key).

Any number can be written as itself raised to the power 1:

$$y = y^1$$

Any number raised to the power 0 is equal to unity:

$$y^0 = 1$$

If two different powers of the same number are multiplied, the product is obtained by summing the powers:

$$y^a \times y^b = y^{a+b}$$

This method can also be used for division using the negative exponent notation:

$$\frac{y^a}{y^b} = y^a \times y^{-b} = y^{a-b}$$

Exercise

Evaluate the following using the various rules described above:

(i) $\qquad 2^{1/2} \times 2^{3/2} =$

(ii) $\qquad 2^{0.15} \times 2^{3.7} =$

(iii) $\qquad \dfrac{2^{1/2}}{2^{3/2}} =$

(iv) $\qquad \dfrac{2^{3/2}}{2^{0.75} \times 2^{0.75}} =$

(Answers: (i) $2^2 = 4$, (ii) $2^{3.85} = 14.42$; (iii) $2^{-1} = \frac{1}{2}$; (iv) $2^0 = 1$.)

Note

The last result shows that:

$$y^n \times y^{-n} = 1$$

Some other 'interesting facts' are worth gathering together:

If y is greater than 1, then y^n is greater than unity for any positive exponent, becoming larger as n increases and tending to unity as n tends to zero.

If y is less than unity, then y^n is less than unity for any positive exponent, tending to zero as n increases and tending to unity as n tends to zero.

If y is greater than unity, then y^{-n} is positive but less than unity for any positive n, tending to zero as n increases and tending to unity as n tends to zero.

If y is less than unity, then y^{-n} is positive and greater than unity for any positive n, becoming larger as n increases and tending to unity as n tends to zero.

Calculators will generally find y^n directly of a negative number ($y < 0$) if n is an integer, but usually fail with negative numbers with general exponents.

2.3 Exponentials

We have seen that any number can be raised to any power, but there are two particularly special cases.

Much of our thinking is based on the powers of 10 (the *decadic scale*) as discussed in section 2.1.

The second 'special' number in this context is that represented by the symbol e which we will find frequently in chemistry, e.g. in kinetics or radio-activity.

The number e itself is *irrational*, i.e. like π it cannot be exactly represented by a finite number of decimal places. To 25 significant figures, the value of e (or e^1) is

$$e = 2.718\ 281\ 828\ 459\ 045\ 235\ 360\ 287$$

Entering 1 and using the e^x button on a calculator yields the above result to fewer significant figures.

Exercises

Using the e^x button on a calculator, evaluate the following:

(i) $\quad e^2$

(ii) $\quad e^{-3.5}$

(Answers: (i) 7.3890561; (ii) 0.03019738.)

The concentration of a chemical A in a reaction follows *first order decay* governed by the equation
$$[A] = [A]_0\,e^{-kt}$$

If the initial concentration $[A] = 0.1$ mol dm^{-3}, calculate the concentration after 50 s if $k = 0.01$ s^{-1}

$$[A] =$$

(Answer: substitute to give $[A] = 0.1\,\text{mol dm}^{-3} \times e^{-0.01s^{-1} \times 50s} = 0.1\,\text{mol dm}^{-3} \times e^{-0.5} = 0.06065\,\text{mol dm}^{-3}$.)

All of the general rules for exponentials operate with e^x, and these are summarised in the table below:

$e^x > 0$	for any x	
$e^x > 1$	for $x > 0$	see graph on p. 37
$0 < e^{-x} < 1$		
$e^{-x} = \dfrac{1}{e^x}$	negative exponents	
$e^a e^b = e^{a+b}$		
$\dfrac{e^a}{e^b} = e^{a-b}$	additivity of exponents	
$\left(e^x\right)^2 = e^{2x}$		
$e^0 = 1$		
$e^{-\infty} = 0$	special values	
$e^{+\infty} = \infty$		

Exercise

This exercise combines three of the points discussed so far.

Calculate the value of a rate constant k from the Arrhenius formula

$$k = Ae^{-E/RT}$$

at $T = 500$ K if $A = 10^{10}$ s^{-1}, $E = 70$ kJ mol^{-1} and $R = 8.314$ J K^{-1} mol^{-1}.

$$k =$$

Hints: remember the rule for entering a number such as 10^{10} into a calculator; note that E is quoted in terms of kJ whilst R is in J K^{-1} mol^{-1}.

(Answer: $k = 1 \times 10^{10}$ s^{-1} e$^{-70000/8.314 \times 500} = 1 \times 10^{10}$ s^{-1} e$^{-16.84} = 486.3$ s^{-1}.)

The definition of e^x involves an *infinite series*:
$$e^x = 1 + x + \frac{1}{2}x^2 + \frac{1}{3!}x^3 + ... \frac{1}{n!}x^n ...$$

The notation 3! and n! refers to the *factorial*:

$$3! = 3 \times 2 \times 1 = 6;$$
$$7! = 7 \times 6 \times 5 \times 4 \times 3 \times 2 \times 1 = 5040$$
$$n! = n \times (n-1) \times (n-2) \times .... \times 2 \times 1$$

The expansion above is particularly useful for evaluating e^x for small values of x, when only the first few terms need be retained.

Exercise

Compare the values of e^x with $1 + x$ for the following:

(i) $x = 0.5$ $e^x =$ $1 + x =$

(ii) $x = 0.1$ $e^x =$ $1 + x =$

(iii) $x = 0.01$ $e^x =$ $1 + x =$

(iv) $x = 10^{-4}$ $e^x = 1.00010000\overline{5}$ $1 + x = 1.0001$ $\therefore e^x$

The expansion for e^{-x} can be written as

$$e^{-x} = 1 - x + \frac{1}{2}x^2 - \frac{1}{3!}x^3 + ... \frac{(-1)^n}{n!}x^n ...$$

so for small x, $e^{-x} \approx 1 - x$

Exercise

The term $1 - e^{-E/RT}$ occurs in statistical thermodynamics. If $E \ll RT$, the exponent E/RT will be small. Use the above approximate form to find how $1 - e^{-E/RT}$ behaves at high temperature. (*Hint*: write E/RT as x, replace e^{-x} by $1 - x$ and then subtract this from 1.)

(Answer: $1 - e^{-E/RT} \approx E/RT$ at large T.)

The basic form for the functions e^x and e^{-x} are shown in the graph below:

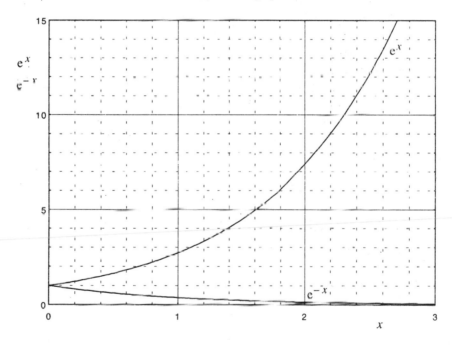

Exponential decay occurs widely in chemical kinetics with the concentration of a reactant A decreasing according to the equation:

$$[A] = [A]_0 e^{-kt}$$

as described earlier. Here $[A]_0$ is the initial concentration and k the rate constant.

The larger the value of k, the faster the rate of decay, as indicated in the graph below:

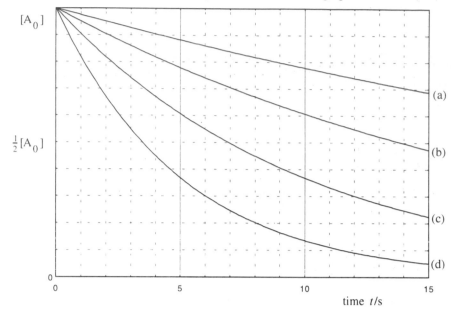

Curves (a) – (d) correspond to the following values of k: (a) $k = 0.025$ s^{-1}; (b) $k = 0.05$ s^{-1}; (c) $k = 0.1$ s^{-1}; (d) $k = 0.2$ s^{-1}

Exercise

From the graph, estimate the first 'half-life' of the reaction, i.e. the time taken for [A] to fall to one half of its initial concentration [A]$_0$, for each of curves (b), (c) and (d)

curve (b), $k = 0.05$ s^{-1} $t_{1/2} =$

curve (c), $k = 0.1$ s^{-1} $t_{1/2} =$

curve (d), $k = 0.2$ s^{-1} $t_{1/2} =$

(Answers: (b) $t_{1/2} = 13.9$ s, (c) $t_{1/2} = 6.9$ s; (d) $t_{1/2} = 3.5$ s.)

Exercise

For curves (c) and (d) estimate the 'second half-life', i.e. the time taken for [A] to fall from $\frac{1}{2}[A]_0$ to $\frac{1}{4}[A]_0$: for curve (d) also estimate the third half-life in which [A] falls from $\frac{1}{4}[A]_0$ to $\frac{1}{8}[A]_0$.

curve (c)   second half-life =

curve (d)   second half-life =
                   third half-life =

(Answers: curve (c) 6.9 s; curve (d) 3.5 s and 3.5 s.)

Notes

It is a characteristic feature of exponential decay that all half-lives are the same for a given value of k. This also means that the half-life is independent of the initial concentration [A]$_0$.

We will return to these points in the next section when we consider logarithms.

2.4   Logarithms

Logarithms arise widely in chemistry:

they form the basis for pH and related measures of concentration;
they are used in graphs to obtain straight-line plots or to plot data over a large range.
they provide the *inverse function* for exponentials;
they arise naturally from *integration*;

Two 'types' of logarithm are particularly important:

logarithms to the base 10	lg, or $\log_{10}$
natural logarithms, i.e. logarithms to the base e	ln

The logarithm of a number x is the power to which the *base* must be raised to give that number:

Example

the logarithm to the base 10 of 100 is 2,

$$\text{i.e. } \log_{10}(100) = 2, \text{ because } 10^2 = 100$$

the natural logarithm of 100 is 4.60517,

$$\text{i.e. } \ln(100) = 4.60517, \text{ because } e^{4.60517} = 100$$

Exercise

Evaluate the logarithm to the base 10 of

(i)   1 $\log_{10}(1) =$

(ii)   25 $\log_{10}(25) =$

(iii)   0.01 $\log_{10}(0.01) =$

Evaluate the natural logarithm of

(iv)   1 $\ln(1) =$

(v)   25 $\ln(25) =$

(vi)   0.01 $\ln(0.01) =$

(Answers: (i) 0, (ii) 1.39794, (iii) −2; (iv) 0, (v) 3.21888, (vi) −4.60517.)

The following are true for any type of logarithm:

The logarithm of 1 is 0.

The logarithm of a number less than 1 is negative.

The logarithm of a number greater than 1 is positive.

The logarithm of a negative number is not defined.

The logarithm of $1/x$ has the same magnitude but opposite sign to the logarithm of x.

The following table summarises the important points of logarithms (note that the same rules apply to both logarithms to the base e (left-hand column) and to the base 10 (right-hand column)

$\ln(e^x) = x$	$\log_{10}(10^x) = x$
$\ln(x^a) = a\ln(x)$	$\log_{10}(x^a) = a\log_{10}(x)$
$\ln(xy) = \ln(x) + \ln(y)$	$\log_{10}(xy) = \log_{10}(x) + \log_{10}(y)$
$\ln(x/y) = \ln(x) - \ln(y)$	$\log_{10}(x/y) = \log_{10}(x) - \log_{10}(y)$
$\ln(1/x) = -\ln(x)$	$\log_{10}(1/x) = -\log_{10}(x)$
$\ln(x) = 2.303\,\log_{10}(x)$	$\log_{10}(x) = 0.434\,\ln(x)$

(note: $2.303 = \ln(10)$ and $0.434 = \log_{10}(e)$)

Worked Example

$\ln(3) = 1.0986$, calculate $\ln(9)$.

As $9 = 3^2$, we can use the form $\ln(x^a) = a\ln(x)$.

$\ln(9) = \ln(3^2) = 2\ln(3) = 2 \times 1.0986 = 2.1972$.

Exercises

$\ln(3) = 1.0986$ and $\ln(2) = 0.6931$. From this information, calculate:

(i) $\ln(6) =$

(ii) $\ln(\frac{1}{2}) =$

(iii) $\ln(\frac{1}{4}) =$

(iv) $\log_{10}(\frac{1}{4}) =$

Calculate

(v) $\log_{10}(0.1) =$

(vi) $\log_{10}(10^{-7}) =$

(Answers: (i) $\ln(6) = \ln(2) + \ln(3) = 1.7917$; (ii) $\ln(\frac{1}{2}) = -\ln(2) = -0.6931$; (iii) $\ln(\frac{1}{4}) = \ln((\frac{1}{2})^2) = 2\ln(\frac{1}{2}) = -1.3862$; (iv) $\log_{10}(\frac{1}{4}) = 0.434\ln(\frac{1}{4}) = -0.6016$; (v) -1; (iv) -7.)

The application of ln as the inverse of the exponential function, $\ln(e^x) = x$, is illustrated in the next example·

Worked Example

The exponential decay examined in the previous section $[A] = [A]_0 e^{-kt}$ can be recast by taking natural logarithms of both sides:

$$\ln[A] = \ln\left\{[A]_0 e^{-kt}\right\} = \ln[A]_0 + \ln\left(e^{-kt}\right)$$

so

$$\ln[A] = \ln[A]_0 - kt$$

One significance of this is that a plot of ln[A] versus t should thus be a straight-line with a gradient of $-k$ and intercept of $\ln[A]_0$ (remember the equation for a straight line is $y = mx + c$ where m is the gradient and c the intercept: see section 3.1 for a discussion of straight-line graphs). In experiments, therefore, we would plot ln[A] against time to find the *reaction rate constant, k*.

We can also use this approach to relate the half-life $t_{1/2}$ to the rate constant k:

Exercise

At the end of the first half-life, $t = t_{1/2}$ and $[A] = \frac{1}{2}[A]_0$

Substitute these into the equation $[A] = [A]_0 e^{-kt}$

$$\frac{1}{2}[A]_0 = [A]_0 e^{-kt_{1/2}}$$

Cancel the initial concentration, as appropriate, and then take natural logarithms of each side. Rearrange to give $t_{1/2}$ in terms of k.

$$.5 = e^{-kt_{1/2}}$$
$$\ln .5 = -kt_{1/2}$$
$$t_{1/2} = -\frac{\ln .5}{k}$$
$$= \ln \frac{1}{.5} \times \frac{1}{k}$$
$$= \ln 2 \times \frac{1}{k}$$
$$= \frac{\ln 2}{k}$$

$$1 = 2e^{-kt_{1/2}}$$
$$\ln 1 = \ln 2 - kt_{1/2}$$
$$kt_{1/2} = \ln 2 - \ln 1 = \ln 2$$
$$t_{1/2} = \frac{\ln 2}{k}$$

Check your answer by comparing the predicted half-life for a system with $k = 0.1$ s^{-1} with your estimate from curve (c) in the previous section.

$$t_{1/2} =$$

(Answer: $t_{1/2} = \ln(2)/k$: $t_{1/2} = 6.93$ s.)

$-\log .5 = +.3$

$\log 2 = .3$

BECAUSE $5 - $

$\frac{1}{5^{-1}} = \left(\frac{1}{5}\right)^{-1}$

so $\log 5 = -\log\left(\frac{1}{5}\right)$ or $-\log 5 = \log\left(\frac{1}{5}\right)$

Exercise

The Arrhenius equation for the temperature dependence of a reaction rate constant has the form $k = Ae^{-E/RT}$ where A is the pre-exponential factor, E the activation energy and R the universal gas constant.

Re-express this equation in (natural) logarithmic form:

$\ln(k) =$

(Answer: $\ln(k) = \ln(A) - \dfrac{E}{RT}$)

This form will also be relevant when we come to discuss straight-line plots

The pH scale is defined by the equation:

$$pH = -\log_{10}\left[H^+\right]$$

with the concentration quoted in M (i.e. mol dm^{-3}). (More precisely, we should take the logarithm of the *activity* of the hydrogen ion a_{H^+}, which has no units: for dilute solutions $a_{H^+} \approx [H^+]/M$.)

This scale is particularly useful as it is natural for us to work in concentrations that are simple powers of 10, e.g. 1M, 0.1M, 0.01M etc.

A dilution by a factor of 10 corresponds to a change of 1 pH unit (for a strong acid that is fully ionised).

The negative sign means that pH's are usually positive even though solutions are usually less than 1 molar.

Exercise

Calculate the pH of the following solutions of strong acids:

(i) 10^{-3} M HCl   pH =

(ii) 10^{-4} M HCl   pH =

(iii) 10^{-4} M H$_2$SO$_4$ pH =

(iv)   2M H$_2$SO$_4$ pH =

(Answers: (i) pH = 3; (ii) pH = 4; (iii) pH = 3.699; (iv) pH = -1.386.)

High pH values corresponds to low hydrogen ion concentrations: negative values correspond to highly acidic solutions, in excess of 1 M.

2.5   Exponentials and 10^x as 'antilogarithms'

ie Antilogs

Just as ln and $\log_{10}$ are the *inverse functions* of e^x and 10^x, so the converse is true, with e^x and 10^x providing the *antilogarithm* function.

$x = 10^2 = 100$

This relies on the facts that

$\log_{10} 10^2 = 2$ $10^{\log_{10} x} = x$

$$e^{\ln(x)} = x \qquad \text{and} \qquad 10^{\log_{10}(x)} = x$$

$10^2 = 100$
OK

Worked Example

$e^{\ln x} = x$

If $\ln(x) = 3.7$, evaluate x.

$\ln(x) = 3.7$, therefore taking antilogarithms to the base e:

$$x = e^{3.7} = 40.447$$

Exercises

(i) What is the H^+ concentration in a solution of pH = 5.3?

(ii) The intercept c from a plot of $\ln[A]$ versus time, where $[A]$ is the concentration of a reactant A, corresponds to $\ln[A]_0$, where $[A]_0$ is the initial concentration.

$\ln[A]$

$c = \ln[A]_0$

t

If $c = -6.91$, calculate $[A]_0$.

$\ln[A]_0 = -6.9$

$[A]_0 = 10^{-3}$

(Answers: (i) $-\log_{10}[H^+] = 5.3$, therefore, $[H^+] = 10^{-5.3}$ M = 5.01×10^{-6} M;
(ii) $[A]_0 = e^{-6.91} = 1 \times 10^{-3}$ M.)

2.6 Miscellaneous Practice

(i) $10^3/10^5 =$

(ii)   Simplify $\ln(x/y^3)$

(iii)   Expand the following

$$\ln\left\{\frac{2\pi mkT}{h^2}\right\}^{3/2} =$$

(iv)   If $A = 3 \times 10^{12}\ \text{s}^{-1}$ and $E = 120\ \text{kJ mol}^{-1}$, calculate the value of the rate constant at 600 K

         $k =$

(v)   Without using a calculator or tables, evaluate

$$\ln(e^6) + \log_{10}(10^{3.2}) - e^{\ln(2.4)} + 10^{\log_{10}(1/5)}$$

[handwritten: $6 + 3.2 - 2.4 + .2 = 7$]

[handwritten: $\log_{10} .2 = -.699$; $10^{-.699}$]

(vi)   Calculate the pH of a 0.05 M solution of HNO_3

(vii)   A solution is quoted as having a pOH of 4.2. Calculate the corresponding OH^- concentration:

(Answers: (i) $10^{-2} = 0.01$, (ii) $\ln(x) - 3\ln(y)$; (iii) $\frac{3}{2}\ln(2) + \frac{3}{2}\ln(\pi) + \frac{3}{2}\ln(m) + \frac{3}{2}\ln(k) + \frac{3}{2}\ln(T) - 3\ln(h)$; (iv) $k = 107\ \text{s}^{-1}$; (v) $6 + 3.2 - 2.4 - 0.2 = 6.6$; (vi) pH = 1.30; (vii) pOH $= -\log_{10}[OH^-]$, so $[OH^-] = 6.31 \times 10^{-5}$M.)

Summary of Section

The material in this section has covered the basic features of powers, exponentials and logarithms as they arise in chemistry. The following should be familiar:

- manipulating powers of 10 and SI prefixes

- squares, square roots, cubes and other powers and roots

- *e*, the exponential function and exponential growth and decay

- logarithms, both natural and to the base 10

$$e^{-x} = 1/e^x \qquad \ln(e^x) = x \qquad \log_{10}(10^x) = x \qquad 10^{-x} = 1/10^x$$

$$e^a \times e^b = e^{a+b} \qquad \ln(x^a) = a\ln(x) \qquad \log_{10}(x^a) = a\log_{10}(x) \qquad 10^a \times 10^b = 10^{a+b}$$
$$e^a/e^b = e^{a-b} \qquad\qquad\qquad\qquad\qquad\qquad\qquad\qquad 10^a/10^b = 10^{a-b}$$

$$\ln(xy) = \ln(x) + \ln(y) \qquad \log_{10}(xy) = \log_{10}(x) + \log_{10}(y)$$
$$1 < e^x \qquad \ln(x/y) = \ln(x) - \ln(y) \qquad \log_{10}(x/y) = \log_{10}(x) - \log_{10}(y) \qquad 1 < 10^x$$
$$0 < e^{-x} < 1 \qquad\qquad\qquad\qquad\qquad\qquad\qquad\qquad\qquad\qquad 0 < 10^{-x} < 1$$

$$\ln(1/x) = -\ln(x) \qquad \log_{10}(1/x) = -\log_{10}(x)$$

$$e^0 = 1 \qquad\qquad\qquad\qquad\qquad\qquad\qquad\qquad\qquad 10^0 = 1$$
$$e^{-\infty} = 0 \qquad \ln(x) = 2.303\log_{10}(x) \qquad \log_{10}(x) = 0.434\ln(x)$$

SECTION 3

Calculus 1: Differentiation

Differentiation arises naturally whenever we consider the rate at which change occurs: the *derivative* of one quantity with respect to another tells us how and how quickly that quantity varies as the other is changed. For example, we may be interested in how quickly a particular concentration changes with time in a reaction or how the internal energy of a system varies with the pressure or temperature. We will see that the derivative can be identified with the *slope* of a graph of one quantity against the other.

Calculus 1: Differentiation

3.1 Straight-line graphs

The equation: $\qquad y = m\,x + c$

describes a straight line if y is plotted against x. The gradient of the graph will have the value m and the line will intersect the y-axis at the value c

$$y = m\,x + c$$

$$\text{gradient} = m$$
$$\text{intercept} = c$$

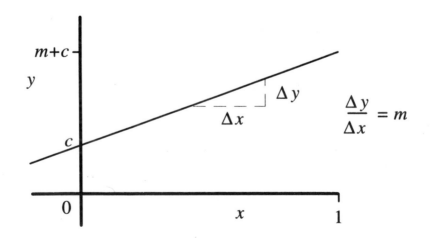

We can show that the gradient is given by m, and illustrate a general method to be used throughout this section, in the following way:

Let the value of y at $x = x_1$ be y_1, then

$$y_1 = m x_1 + c$$

If y similarly has the value y_2 when $x = x_2$, then

$$y_2 = m x_2 + c$$

As illustrated in the figure above, we can evaluate the gradient by dividing the *increment* in y, Δy (sometimes known as the *rise*), by the corresponding *increment* in x, Δx (known as the *run*).

If we take $\Delta y = y_2 - y_1$, then $\Delta x = x_2 - x_1$.

Substituting for y_1 and y_2, we also find that:

$$\Delta y = m x_2 + c - m x_1 - c = m (x_2 - x_1) = m \Delta x$$

so

$$\text{gradient} = \frac{rise}{run} = \frac{\Delta y}{\Delta x} = \frac{m \Delta x}{\Delta x} = m$$

The general method is, then, find expressions for the increments Δy and Δx; substitute for y in terms of x from the original equation; find the ratio $\Delta y/\Delta x$ which is equal to the gradient

Exercise

Sketch the line $y = 3x + 2$ and show, using the above method, that the gradient is 3.

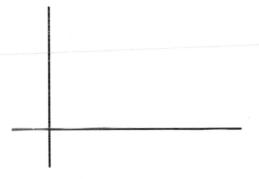

$y_2 =$

$y_1 =$

$\Delta y = y_2 - y_1 =$

$\Delta x =$

$m = \Delta y/\Delta x =$

Exercise

What is the gradient of the line $y = c$, where c is a constant?

(Answer: $\Delta y = c - c = 0$, $\therefore \Delta y/\Delta x = 0$: gradient $= 0$.)

Note This latter exercise simply deals with a special case of the general form $y = m x + c$, with $m = 0$.

3.2   Nonlinear graphs

The straight-line graph has one particularly special feature: the gradient of the line is the same everywhere along the line — it does not vary with x or y.

In general, a plot of one physical quantity against another in chemistry may give rise to a curve. On such a curve, the gradient varies with position.

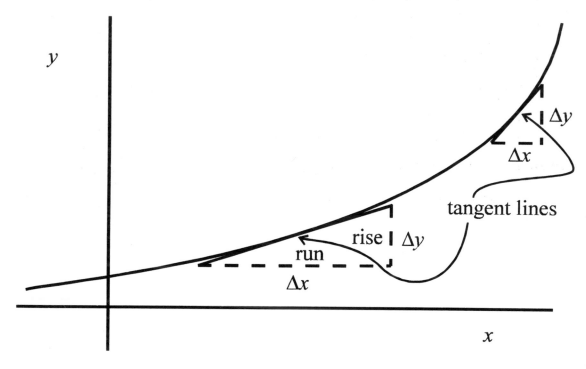

The Δy and Δx constructions are shown at two places 'along' the curve; each has the same Δy, but Δx is clearly different in the two cases (being smaller along the steeper section of the curve).

Also indicated are the *tangent lines*, which are straight lines drawn so as to touch the curve at a given point and so as to have the same slope, $\Delta y/\Delta x$, as the curve at that point.

The tangent line gives the *local slope* at the given point

One method of evaluating the slope of a curve at a given point, then, is to draw the tangent, measure the corresponding increments in x and y and then calculate the gradient of the tangent.

Such a 'mechanical' approach is sometimes used in connection with experimental data. However, if the equation for y in terms of x along the curve is known, a different approach based on the *calculus of differentiation* is much more versatile.

Worked Example

We can illustrate the general approach with reference to the so-called quadratic curve

$$y = x^2$$

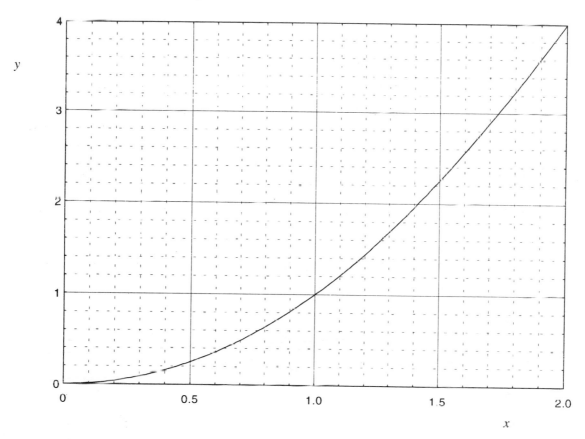

First, use the tangent-drawing method to estimate the gradient at some point, say at $x = 1$ (when $y = 1^2 = 1$).

Draw a tangent to the curve at the point (1, 1) and by measuring the run and rise Δx and Δy, estimate the gradient $m = \Delta y / \Delta x$

$$m =$$

Rather than drawing a tangent, we might approach this task in a different way. If we calculate the values of y at two points x_1 and x_2 from the equation, we can then determine values for Δx and Δy from these.

If we choose $\qquad x_1 = 1,$ $\qquad$ then $\qquad y_1 = 1^2 = 1$

and if we take $\quad x_2 = 1.5,$ $\qquad$ then $\qquad y_2 = 1.5^2 = 2.25$

Thus, $\qquad\qquad\qquad\qquad \Delta x = 1.5 - 1 = 0.5$
$$\Delta y = 2.25 - 1 = 1.25$$

and

$$m = \frac{\Delta y}{\Delta x} = \frac{1.25}{0.5} = 2.5$$

This calculation actually gives us the gradient of a chord connecting the points (1, 1) and (1.5, 2.25).

Draw this chord onto your graph. It is reasonably clear that this chord is steeper than the tangent to (1, 1) drawn previously.

A better estimate of the gradient would be obtained if the second point were chosen closer to the point of interest:

thus, we might choose

$$x_2 = 1.1 \qquad \text{so} \qquad y_2 = 1.1^2 = 1.21$$

with $x_1 = 1$ and $y_1 = 1$ as before.

Now $\Delta x = 1.1 - 1 = 0.1$ and $\Delta y = 1.21 - 1 = 0.21$,

$$\text{giving } m = 0.21/0.1 = 2.1.$$

Exercise

Repeat this calculation for other choices of x_2, with x_2 being chosen to lie closer and closer to x_1. Enter your results in the table below, retaining as many significant figures in each column as possible:

---

x_1	y_1	x_2	$y_2 = x_2^2$	$\Delta x = x_2 - x_1$	$\Delta y = y_2 - y_1$	$m = \dfrac{\Delta y}{\Delta x}$
1	1	1.05				
1	1	1.01				
1	1	1.001				
1	1	1.0001				
1	1	1.00001				

From the trend in the last column, estimate the *limiting value* of the slope m as x_2 tends to x_1, i.e. as the increment Δx becomes *infinitesimally small*:

$$\lim_{\Delta x \to 0} (m) = $$

Compare this result with your estimate from drawing the tangent and measuring Δx and Δy:

The results in the table strongly suggest that the gradient m tends to the value 2 at the point (1, 1) as we allow the increment to become infinitesimally small.

We can also show this from the equation, and hence obtain a general expression for the gradient of this curve at any value of x.

This is an example of *infinitesimal calculus* leading to the *differentiation* of the function $y = x^2$.

The function of interest is
$$y = x^2$$

If x increases by a small amount Δx from x to $x + \Delta x$, then the corresponding increase in y, to $y + \Delta y$ can be calculated using this equation:

$$y + \Delta y = (x + \Delta x)^2 = (x + \Delta x)(x + \Delta x)$$

If we now multiply out the brackets on the right-hand side we obtain

$$y + \Delta y = x^2 + 2x\,\Delta x + (\Delta x)^2$$

To find the change in y, we must subtract away the original value, i.e. subtract y from the left-hand side and, correspondingly, subtract x^2 from the right-hand side, so

$$\Delta y = 2x\,\Delta x + (\Delta x)^2$$

The gradient of the curve m is given by $\Delta y/\Delta x$, so dividing through by Δx we have:

$$m = \frac{\Delta y}{\Delta x} = 2x + \Delta x$$

If we now allow the increment Δx to become vanishingly small, the second term on the right-hand side (i.e. Δx itself) tends to zero. The $2x$ term remains non-zero.

- The *quotient* $\Delta y/\Delta x$ is a ratio of two terms that are both tending to zero, and the above equation tells us that this quotient tends to $2x$ as Δx tends to zero.

- In order to indicate that we have taken this limiting process, we replace the symbol Δ (which is used for a finite — small but not that small — change) by d (which is used for an infinitesimal change in x or y). We thus obtain the *differential* form

$$m = \frac{\mathrm{d}y}{\mathrm{d}x} = 2x$$

The notation dy/dx is used to denote the *derivative of y with respect to x* and tells us how quickly y changes in response to a variation in x.

The derivative of the equation $y = x^2$ is dy/dx = $2x$.

Thus the slope at any point is given by $2x$.

For $x = 1$, dy/dx = 2, which is the result we obtained by our earlier methods, but now we can calculate the slope for any x.

Exercise

Follow the same analysis method to obtain the derivative dy/dx for $y = x^3$.

- You will need to use the expanded form (see section 1.1)

$$(x + \Delta x)^3 = x^3 + 3x^2 \Delta x + 3x(\Delta x)^2 + (\Delta x)^3$$

(Answer: $y + \Delta y$ = expanded form above; subtract y from left-hand side and x^3 from right-hand side and divide through by Δx to give

$$\Delta y/\Delta x = 3x^2 + 3x\Delta x + (\Delta x)^2$$

As $\Delta x \to 0$, the final two terms vanish, yielding the derivative $dy/dx = 3x^2$.)

3.3   Differentiation of other functions

The general form for representing that *y is a function of x* (i.e. the value of y depends on the value chosen for x) is to write $y = f(x)$

In the examples considered above in sections 3.1 and 3.2 we have taken $f(x) = mx + c$, $f(x) = x^2$ and $f(x) = x^3$ respectively.

For any function $f(x)$, the derivative dy/dx gives the gradient of the curve when the function is plotted against x.

In some cases, the derivative is also a (normally different) function of x, e.g. for $y = f(x) = x^2$, $dy/dx = 2x$, indicating simply that the gradient depends on x and so we have a curve not a straight line.

The table below indicates the derivative of a number of functions that arise in chemistry.

Once we have learnt to use this table, and a few additional rules, we will have a toolkit for differentiating virtually any function that arises. (The main trick remaining is to identify y and x in terms of our chemical equation: in a kinetics problem for instance we may wish to find the rate at which a concentration C varies with time t, so that $y = C$ and $x = t$ in the notation below.)

$y = f(x)$	dy/dx
c	0
$mx + c$	m
x^2	$2x$
x^3	$3x^2$
x^n	$n\, x^{n-1}$
e^x	e^x
e^{ax}	$a\, e^{ax}$
$\ln(x)$	$1/x$
$\sin(x)$	$\cos(x)$
$\cos(x)$	$-\sin(x)$
$\sin(ax)$	$a\cos(ax)$
$\cos(ax)$	$-a\sin(ax)$

(handwritten note in margin: NB $\tan x = \dfrac{\sin x}{\cos x}$ *)*

The derivative of a constant is zero

The derivative of a general n-th order *polynomial* term x^n gives $n\, x^{n-1}$, so the order is reduced by one. This form applies even if n is not an integer.

The derivative of the exponential function e^x is equal to the function itself.

The derivative of the natural logarithm function yields $1/x$.

The trigonometric functions are related through their derivatives.

Exercises

Differentiate the following with respect to x:

(i) $y = x^4$ $dy/dx =$

(ii) $y = e^{3x}$ $dy/dx =$

(iii) $y = 1/x^2$ $dy/dx =$ (note $1/x^2 = x^{-2}$)

(iv) $y = \cos(3\pi x)$ $dy/dx =$

(v) $y = e^{-2x}$ $dy/dx =$

(Answers: (i) $4x^3$; (ii) $3e^{3x}$; (iii) $-2x^{-3} = -2/x^3$; $-3\pi\sin(3\pi x)$; (v) $-2e^{-2x}$.)

In order to apply these results to a wider range of functions we need to know the following additional rules:

(i) differentiation of a sum

If the function $f(x)$ can be broken down into a sum of other functions, then each can be differentiated separately and the derivative of $f(x)$ is then the sum of these pieces:

Example

If

$$y = x^3 + x^2 - x + 1$$

then

$$dy/dx = 3x^2 + 2x - 1$$

This rule also implies that

if

$$f(x) = ax^n \qquad dy/dx = a\,n\,x^{n-1}$$

so multiplying the function by a constant a simply multiplies the gradient by the same factor.

e.g.   if

$$y = 3x^3 + 4x^2 - 2x + 1$$

then

$$dy/dx = 9x^2 + 8x - 2$$

(ii) differentiation of a product

If the function $f(x)$ can be written as the product of two other functions, $u(x)$ and $v(x)$, then the derivative of $f(x)$ can be obtained from the following rule:

$$y = f(x) = u(x) \cdot v(x)$$

then

$$\frac{dy}{dx} = \frac{d(uv)}{dx} = u\frac{dv}{dx} + v\frac{du}{dx}$$

Example

if $y = x^2 e^{3x}$, we can assign $u = x^2$ and $v = e^{3x}$, so

$$\frac{dy}{dx} = x^2 \frac{d(e^{3x})}{dx} + e^{3x}\frac{d(x^2)}{dx} = 3x^2 e^{3x} + 2xe^{3x} = (3x^2 + 2x)e^{3x}$$

(iii) differentiation of a quotient

A slight variation on the above

If $f(x)$ can be expressed as a quotient of two functions, $f(x) = u(x)/v(x)$.

Then
$$\frac{dy}{dx} = \frac{d(u/v)}{dx} = \frac{v\dfrac{du}{dx} - u\dfrac{dv}{dx}}{v^2}$$

Example

(handwritten annotations:) $\dfrac{\cos x \frac{d\sin x}{dx} - \sin x \frac{d\cos x}{dx}}{\cos^2 x} = \dfrac{\cos x \cos x - \sin x(-\sin x)}{\cos^2 x} = \dfrac{\cos^2 x + \sin^2 x}{\cos^2 x}$

If $y = \sin(x)/\cos(x)$, then taking $u = \sin(x)$ and $v = \cos(x)$ we get:

(handwritten: wrong way round)

$$\frac{dy}{dx} = \frac{\sin(x)\dfrac{d\cos(x)}{dx} - \cos(x)\dfrac{d\sin(x)}{dx}}{\cos^2(x)} = \frac{-\sin(x)\sin(x) - \cos(x)\cos(x)}{\cos^2 x} = -\frac{\sin^2(x) + \cos^2(x)}{\cos^2(x)} = \frac{-1}{\cos^2(x)}$$

The final simplification in this result comes from the fact that $\sin^2(x) + \cos^2(x) = 1$.

This result is the derivative of $\tan(x)$ since $\tan(x) = \sin(x)/\cos(x)$.

(iv) the chain rule for the differentiation of a function of a function

This rule is useful for functions such as $y = f(x) = e^{ax^2}$, where the exponent itself is a function of x.

If the function $f(x)$ can be rewritten as a function of u, where u is also a function of x, often written as

$$y = f(u(x))$$

then it is often simpler to split the differentiation into two parts according to the chain rule formula:

$$\frac{dy}{dx} = \frac{dy}{du} \times \frac{du}{dx}$$

Example

Considering $y = f(x) = e^{ax^2}$, we can write this as $y = e^u$ with $u = ax^2$. Then

$$\frac{dy}{dx} = \frac{d(e^u)}{du} \times \frac{d(ax^2)}{dx} = e^u \times 2ax = 2ax\, e^{ax^2}$$

where u has been replaced by ax^2 in the exponential to give the final result solely in terms of x.

Exercises

Differentiate the following:

(i) $y = \cos(3x) - \sin(4x) + 7x^5$

(ii) $y = \sin(2x)\cos(3x)$

(iii) $y = 3x^2 \ln(x)$

(iv) $y = \ln(3x)$

(v) $y = Ae^{-B/x}$

(vi) $y = \ln(x^3)$

(vii) $y = xe^{-x}$

(Answers: (i) $-3\sin(3x) - 4\cos(4x) + 35x^4$; (ii) $2\cos(2x)\cos(3x) - 3\sin(2x)\sin(3x)$; (iii) $6x \ln(x) + 3x$; (iv) $1/x$; (v) $(AB/x^2)e^{-B/x}$; (vi) $3/x$; (vi) $(1-x)e^{-x}$.)

3.4 Rates of change

The derivative dy/dx relates to the 'rate of change of y with x', i.e. how y changes as x varies. We can always think of this in terms of the gradient of a graph of y plotted against x.

If dy/dx is positive, y increases as x increases (and decreases as x decreases):

If dy/dx is negative, y decreases as x increases (and increases as x decreases).

If dy/dx is large, then the variation of y with x is rapid;

If dy/dx is small, then the variation of y with x is slow.

Exercise

Mark on the graph shown below, regions where dy/dx is (i) positive and (ii) negative.

Indicate where *y* varies rapidly with *x*.

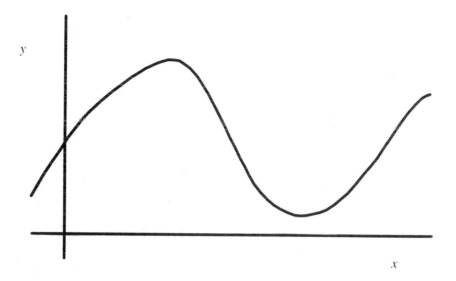

(Answer)

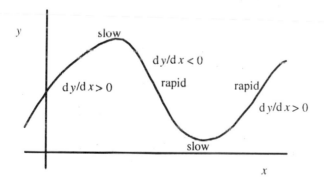

These points are summarised in the table:

magnitude sign	small	large
positive	increasing slowly	increasing rapidly
negative	decreasing slowly	decreasing rapidly

3.5   Finding maxima and minima: higher derivatives

We can use the idea of the derivative as the gradient of a curve to locate a maximum or minimum (also known as an *extremum* or *turning point*) of a function.

The curve above shows both a maximum and a minimum.

To the left of the maximum, y is increasing as x increases, so the gradient is positive

$$\mathrm{d}y/\mathrm{d}x > 0 \text{ before the maximum}$$

To the right of the maximum, y decreases as x increases, so the gradient is negative

$$\mathrm{d}y/\mathrm{d}x < 0 \text{ after the maximum}$$

At the maximum point, the curve is *locally flat*, i.e. exactly at this point the gradient becomes zero:

$$\mathrm{d}y/\mathrm{d}x = 0 \text{ at the maximum}$$

We can use similar arguments about the minimum: to the left the gradient is negative (y decreases as x increases); after the minimum y increases with x (so the gradient is positive): at the minimum point, the curve is locally flat again

$$\mathrm{d}y/\mathrm{d}x = 0 \text{ at the minimum.}$$

We can locate minima or maxima by looking for points at which $\mathrm{d}y/\mathrm{d}x = 0$.

Worked Example

The curve $y = 3x^2 - 2x + 1$ has a minimum. Locate the minimum value of y and the value of x at which this occurs.

In fact, we work the other way round, first finding the value of x at which the minimum occurs and then substituting this into the equation to find the minimum value of y.

First differentiate the equation for y:

$$\mathrm{d}y/\mathrm{d}x = 6x - 2$$

At a minimum, $\mathrm{d}y/\mathrm{d}x = 0$, i.e.

$$6x - 2 = 0$$

so $x = 1/3$ at the minimum point. Substituting this value for x back into the full equation,

$$y_{min} = 3 \times \tfrac{1}{3}^2 - 2 \times \tfrac{1}{3} + 1 = \tfrac{2}{3}$$

The minimum value of y is $\tfrac{2}{3}$ and occurs when $x = \tfrac{1}{3}$.

Exercises

Find the values of x and y at the minimum or maximum in each of the following functions:

(i) $y = 5x^2 - 7x + 9$

(ii) $y = -8x^2 + 4x + 2$

(iii) $y = 3x^3 - 7x^2 + x - 2$

(Answers: (i) $x = 0.7$, $y_{min} = 6.55$; (ii) $x = 0.25$, $y_{max} = 2.5$; (iii) there are two roots to the equation

$dy/dx = 0$, with $x = \dfrac{14 \pm \sqrt{14^2 - 4 \times 9 \times 1}}{2 \times 9} = 1.481$ and 0.075, giving $y = -6.127$ and $y = -1.963$ respectively.)

The recipe, $dy/dx = 0$, applies to both maxima and minima

We can distinguish between a maximum and a minimum by noting that:

for a maximum, dy/dx changes from positive to negative as x increases; whilst

for a minimum, dy/dx changes from negative to positive as x increases.

This difference between maxima and minima refers to how the gradient changes as x varies.

The rate at which the gradient dy/dx changes with x corresponds to the derivative of the gradient with respect to x, which we can write as

$$\frac{d(dy/dx)}{dx} = \frac{d^2y}{dx^2}$$

The notation d^2y/dx^2 is termed *the second derivative of y with respect to x* and corresponds to the rate at which the steepness and sign of the curve of y against x is varying, i.e. to the *curvature* of the graph.

At a maximum, the gradient is decreasing as x increases (as dy/dx is changing from positive to negative); at a minimum, the gradient is increasing with x as dt/dx is changing from negative to positive as x increases:

At a maximum: $dy/dx = 0$ and $d^2y/dx^2 < 0$

At a minimum: $dy/dx = 0$ and $d^2y/dx^2 > 0$.

Worked Example

Show that the turning point in exercise (i) above is a minimum:

The equation is $y = 5x^2 - 7x + 9$, so

$$dy/dx = 10x - 7$$

and

$$d^2y/dx^2 = d(10x - 7)/dx = 10$$

The turning point occurs when $dy/dx = 0$, i.e. at $x = 0.7$. The second derivative is positive for this value of x so this is a minimum.

Note: The second derivative is found simply by treating dy/dx as a function of x and differentiating this with exactly the same rules as before.

In this case, $d^2y/dx^2 = 10$, so the second derivative is constant and the graph has constant curvature.

Exercises

(i) Show that the turning point of the curve $y = -8x^2 + 4x + 2$ is a maximum.

(ii) The curve $y = 3x^3 - 7x^2 + x - 2$ has two turning points. Determine which is a maximum and which is a minimum.

(Answers: (i) $dy/dx = -16x + 4$, $d^2y/dx^2 = -16 < 0$ for all x, so the turning point will be a maximum; (ii) $dy/dx = 9x^2 - 14x + 1$, $d^2y/dx^2 = 18x - 14 = 2(9x-7)$, we have established that the turning points occur at $x = 1.481$ and 0.075, for which $d^2y/dx^2 = 12.658$ and -12.658 respectively, so the first of these, with $y = -6.127$, is a minimum and the second, with $y = -1.963$, is a maximum.)

An example of a second derivative is acceleration:

acceleration (a) is the rate of change of velocity (v) with respect to time, $a = dv/dt$

In turn, velocity is the rate of change of position (s) with time, $v = ds/dt$

so,

$$a = \frac{d}{dt}\left(\frac{ds}{dt}\right) = \frac{d^2s}{dt^2}$$

A car that is 'accelerating rapidly' (i.e. for which a is large and positive) is increasing in speed rapidly: if a is large and negative the speed is falling quickly so the car slows down rapidly.

In addition to second derivatives, we can find the derivative of a function y to any order with respect to x, denoted d^ny/dx^n, by repeatedly differentiating n times.

Exercises

(i) In a particular reaction, the concentration of a reactant [A] decreases with time according to

$$[A] = [A]_0 e^{-kt}$$

where $[A]_0$ is the initial concentration and k is the reaction rate constant. Write an expression for the *rate of reaction*, i.e. the rate at which A is consumed.

(ii) Find the turning point of the function $y = xe^{-x}$ and indicate whether this is a maximum or minimum

(iii) Find the first, second and third derivatives of the function $y = e^x$

$dy/dx =$

$d^2y/dx^2 =$

$d^3y/dx^3 =$

(iv) The distance s of a car from a town is given by $s = 3 + 5t + 2t^2 - 0.1t^3$, where t is the time: write expressions for

(a) the velocity v

(b) the acceleration a

and hence find

(c) the maximum speed attained

(d) the maximum distance from the town achieved

(Answers: (i) $-d[A]/dt = k[A]_0 e^{-kt} = k[A]$; (ii) $dy/dx = (1 - x)e^{-x}$, $\therefore x = 1$ and $y = e^{-1}$ at the turning point, $d^2y/dx^2 = (x - 2)e^{-x} < 0$ for $x = 1$, so this is a maximum; (iii) $dy/dx = d^2y/dx^2 = d^3y/dx^3 = e^x$, all derivatives of e^x are equal to the value of the function itself.; (iv), (a) $v = ds/dt = 5 + 4t - 0.3t^2$; (b) $a = dv/dt = 4 - 0.6t$; (c) we need $dv/dt = a = 0$, so $t = 4/0.6 = 6.667$, giving $v_{max} = 18.33$; (d) we need $ds/dt = v = 0$, so we find $t = (4 + \sqrt{22})/0.6 = 14.5$ (the other root is negative), giving $s_{max} = 191$, at this time, $d^2s/dt^2 = a = -4.7 < 0$, so this is a maximum.)

3.6   Partial differentiation

In thermodynamics and many other situations, there are quantities that depend on more than one variable, e.g. the molar volume V_m depends on both the pressure p and the temperature T, so $V_m = f(p,T)$; the enthalpy H of a system can also be thought of as a function of the pressure and temperature $H = f(p,T)$; many reaction rates depend on more than one concentration.

If we want to ask questions about how such a *multivariate* function changes, we need to be specific about whether one or all of the quantities on which the function depends are allowed to vary or whether only one is.

Example

Consider the function $y = 3x + 4z$;

if x varies by an amount Δx but z is held constant, then the change in y, Δy will be given by $3\Delta x$;

if, however, the variation in x also causes or is accompanied by a change Δz in z, then $\Delta y = 3\Delta x + 4\Delta z$.

Often, it is useful to find how a function changes as one, and only one, of the things it depends on changes, with the other quantities held constant.

This gives rise to the *partial derivative* of the function with respect to the selected quantity. For instance if $y = f(x,z)$, i.e. y is some function of x and z, then:

---

the partial derivative of y with respect to x is denoted $\left(\dfrac{\partial y}{\partial x}\right)_z$

The symbol ∂ ("del") indicates that the differentiation has be performed with other quantities held constant; the subscript z here indicates that there is one quantity, z, that has been held constant.

There will be a corresponding partial derivative $(\partial y/\partial z)_x$ corresponding to the rate at which y changes with z with x held constant.

Exercise

Partial derivatives are used frequently in thermodynamics where functions such as ΔH depend on the *state variables* p and T. Write down the appropriate mathematical representation of partial derivative for (i) the rate of change of ΔH with temperature at constant pressure and (ii) the rate of change of ΔH with pressure at constant temperature

(Answers: (i) $(\partial \Delta H/\partial T)_p$ and (ii) $(\partial \Delta H/\partial p)_T$.)

When performing a partial differentiation, we use exactly the same rules as before, treating the quantities we are not allowing to vary as constants:

Worked Examples

Find the partial derivatives of the function $y = 3xz + 4z$ with respect to each quantity:

$$(\partial y/\partial x)_z = 3z;$$

$$(\partial y/\partial z)_x = 3x + 4$$

Exercise

(a) Calculate $(\partial y/\partial x)_z$ and $(\partial y/\partial z)_x$ for the function $y = 3x + 4z$

(b) Find the partial derivatives of the function $y = 3x^2z^3 + 4z/x$ with respect to each quantity:

(c) The molar volume V_m of an ideal gas is given by the formula

$$V_m = \frac{RT}{p}$$

where T is the temperature and p the pressure. Find the appropriate form for:

(i) $(\partial V_m/\partial T)_p$

(ii) $(\partial V_m/\partial p)_T$

(Answers: (a) $(\partial y/\partial x)_z = 3$, $(\partial y/\partial z)_x = 4$; (b) $(\partial y/\partial x)_z = 6xz^3 - 4z/x^2$; $(\partial y/\partial z)_x = 9x^2z^2 + 4/x$; (c) (i) $(\partial V_m/\partial T)_p = R/p$; (ii) $(\partial V_m/\partial p)_T = -RT/p^2$.)

More than one quantity can be held constant during the partial differentiation process. For example, the potential energy V of an electron in an electric field depends on the three co-ordinates x, y and z: the appropriate notation for the three partial derivatives of V with respect to position are:

$$\left(\frac{\partial V}{\partial x}\right)_{y,z} ; \quad \left(\frac{\partial V}{\partial y}\right)_{x,z} \quad \text{and} \quad \left(\frac{\partial V}{\partial z}\right)_{x,y}$$

In each case, the position along one co-ordinate varies, the position along the other two co-ordinates is held constant.

A familiar analogy for partial differentiation is to consider how altitude varies as we journey across the countryside. We can travel both north-south and east-west. The partial derivative with respect to longitude at constant latitude corresponds to how the height of the landscape above sea-level varies as we move in the east-west direction but keeping the north-south component fixed. The partial derivative with respect to latitude at constant longitude refers to how much we go up or down travelling due north-south without veering to east or west of our current position.

We can maximise a function with respect to a single quantity, at particular fixed values of the other quantities. For instance, if y depends on x and z, we can use the condition $(\partial y/\partial x)_z = 0$ to find a maximum or minimum much as before.

Example

Find the location of the maximum in y with respect to x if $y = 4zx - x^2$

Differentiating with respect to x at constant z gives

$$\left(\partial y / \partial x\right)_z = 4z - 2x$$

For a turning point, we need $(\partial y/\partial x) = 0$, so the condition becomes

$$4z - 2x = 0 \,,\ \text{i.e.}\ \ x = 2z$$

In this case, the position of the turning point in the corresponding value of y will depend on the particular value given to z. This is typically the case in such problems.

If we take $z = \frac{1}{4}$, then the turning point occurs at $x = \frac{1}{2}$, giving $y_{max} = \frac{1}{4}$ for this particular choice of z.

Finally, we can confirm that this is a maximum by finding the second partial derivative of y with respect to x:

$$\left(\partial^2 y / \partial x^2\right)_z = -2$$

which is definitely negative.

In other cases, we can similarly use the condition $(\partial y/\partial z)_x = 0$ to locate a turning point in the z-direction.

The business of finding *global maxima or minima*, i.e. the highest or lowest points on the whole y surface is also of interest in physical chemistry but requires us to deal with *total derivatives*.

The total derivative of a function y has to allow for infinitesimal changes in all of the quantities involved. The notation used here is of the form

$$dy = \left(\frac{\partial y}{\partial x}\right)_z dx + \left(\frac{\partial y}{\partial z}\right)_x dz$$

We read this by saying that the total change dy in y depends on the change dx in quantity x and the change dz in the quantity z. These changes occur in the equation multiplied by the partial derivative terms as indicated because these partial derivatives are the rate at which the function y changes with x and z.

For a global maximum, $dy = 0$, so we need both $(\partial y/\partial x)_z$ and $(\partial y/\partial z)_x$ to be zero simultaneously.

We evaluated a second partial derivative $(\partial^2 y/\partial x^2)_z$ in the previous example. This gave the curvature of the function y in the x direction at some constant value of z.

There are other second derivatives possible which correspond to curvatures in other directions. In fact, if y depends on two quantities x and z, there will be four of these second partial derivatives:

$$\left(\frac{\partial^2 y}{\partial x^2}\right)_z ; \quad \left(\frac{\partial^2 y}{\partial z^2}\right)_x \qquad \text{and} \qquad \left(\frac{\partial^2 y}{\partial x \partial z}\right) ; \quad \left(\frac{\partial^2 y}{\partial z \partial x}\right)$$

$g_{10}10 = 1$

$\log_e e = 1$ i.e $\ln e = 1$

ect to one quantity and then with respect to the other; they
ler with which the two differentiation processes are carried

suggestion using a
from the slope and

s x. We would then

volume V_m with respect to T and p if $V_m = RT/p$

h respect to p to give

site order, i.e. first with respect to p and then with

give the same result.)

ns. Indicate how the

ferentiation process is carried out is unimportant, this

he hydrogen atom are
ctron is falling by the

a reactant decreasing

e plotted graphically. If we choose the appropriate
ip of the above form, then we can hope to use the
quantities. In some cases, even testing whether a
m or disprove a theoretical suggestion.

plotted against time to

ly

lar reaction at several temperatures, we can test
k fits the equation

ept $c = \frac{1}{4}\Re_H$; (b) plot
dient $m = k$.)

$$R = A\, e^{-\frac{E}{RT}}$$
$$\ln k = \ln A - \frac{E}{RT}$$

Exercise

Suggest what function of k might be plotted against what function of T to test this straight line graph. How then might the unknown quantities A and E be determined intercept?

Answer: We can take natural logs of the equation to give

$$\ln(k) = \ln(A) - \frac{E}{R} \times \frac{1}{T}$$

Comparing this with $y = mx + c$, we might choose to plot $\ln(k)$ as y against $1/T$ a obtain $\ln(A)$ as the intercept c and the gradient $m = -E/R$.

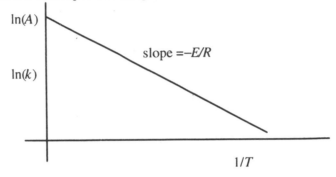

Exercises

Suggest suitable plots to test whether experimental data fit the following equatio unknown quantities can be obtained from the gradient and/or intercept

(a) The wavenumber $\bar{v}$ of lines in the Balmer series in the emission spectrum of t thought to be related to the quantum number n of the energy level from which the ele equation

$$\bar{v} = \Re_H\left(\frac{1}{4} - \frac{1}{n^2}\right)$$

Suggest a suitable plot of the $\bar{v} - n$ data to find the Rydberg constant $\Re_H$.

(b) A reaction is thought to follow first order kinetics, with the concentration of exponentially according to

$$[A] = [A]_0 e^{-kt}$$

where k is the rate constant (which we hope to find). Suggest how $[A]$ may be test for first order behaviour and to allow k to be determined.

(c) A different reaction is thought to follow second order kinetics which would im

$$\frac{1}{[A]} = \frac{1}{[A]_0} + kt$$

What plot would be appropriate here?

(Answers: (a) plot $\bar{v}$ as y against $1/n^2$ as x, the gradient $m = -\Re_H$ and the inter

$\ln[A]$ versus t and obtain k from the gradient $m = -k$; (c) plot $1/[A]$ against t, with gr

In thermodynamics, the information is sometimes presented in a slightly different form. For instance, the dependence of an equilibrium constant K_{eq} on temperature is often expressed via the equation

$$\left(\frac{\partial \ln K_{eq}}{\partial(1/T)} \right)_{p} = -\frac{\Delta H^{\theta}}{R}$$

We can read this as stating that the gradient of a plot of $\ln K_{eq}$ (as y) against $1/T$ (as x) will be equal to $-\Delta H^{\theta}/R$, i.e. this equation is of the form $dy/dx = m$.

Exercise

For an endothermic reaction with $\Delta H^{\theta} > 0$.

(a) How does $\ln K_{eq}$ change as $1/T$ increases?

(b) How does K_{eq} change as T increases?

(c) Does this fit in with your expectations based on le Chatelier's principle?

Answers (a) As the derivative is equal to a negative quantity, $\ln K_{eq}$ decreases as $1/T$ increases, this means (b) that K_{eq} decreases as T decreases or, equivalently K_{eq} increases as T increases (c) this fits the prediction based on le Chatelier that an endothermic reaction proceeds further at higher temperatures.

Summary of Section

The material in this section has introduced the idea of a derivative dy/dx as the slope of a graph of y against x. For straight line graphs, the slope and hence the derivative are constant and so do not depend on x: for other functions, the slope varies as x varies. The following concepts should be familiar:

- slope of a graph

- finding a gradient from the 'rise' Δy and 'run' Δx; $m = \Delta y/\Delta x$

- the idea of a tangent to a curve

- the idea of an infinitesimal change in x or y

- the derivative of simple powers of x, of e^x, of $\ln(x)$ and of $\sin(x)$ and $\cos(x)$

- the product rule $d(uv)/dx = u(dv/dx) + v(du/dx)$

- the chain rule $dy/dx = (dy/du) \times (du/dx)$

- the relationship between the derivative and the rate of change of y

- finding maxima and minima $dy/dx = 0$

- second derivatives such as acceleration as the curvature of a graph

- partial derivatives

- choosing appropriate straight line plots to test experimental data

$y=f(x)$	dy/dx
c	0
$mx + c$	m
x^2	$2x$
x^3	$3x^2$
x^n	$n\, x^{n-1}$
e^x	e^x
e^{ax}	$a\, e^{ax}$
$\ln(x)$	$1/x$
$\sin(x)$	$\cos(x)$
$\cos(x)$	$-\sin(x)$
$\sin(ax)$	$a \cos(ax)$
$\cos(ax)$	$-a \sin(ax)$

SECTION 4

Calculus 2: Integration

Where differentiation is concerned with determining the rate at which quantities change, integration is used to find how much change has occurred as a result of the rate. Integration turns up throughout chemistry but particularly in thermodynamics, reaction kinetics and the kinetic theory of gases.

Calculus 2: Integration

4.1 Interpretation of integration

Integration is the reverse or *'inverse'* process to differentiation. If a function is differentiated and then integrated with respect to the same quantity, we get the original function back again.

Integration is used in various ways in chemistry: in kinetics it is used to calculate the concentration at any given time from the rate of reaction; in thermodynamics it is used to obtain the work done as pressure and volume change or the entropy change as temperature is varied; in the kinetic theory of gases it is used to calculate average velocities.

These applications can be expressed more generally. In the first case, integration is used to find an expression for some quantity of interest if we know the rate at which the quantity is varying, i.e. the slope of the graph. Mathematically, we know dy/dx and use this to find how y depends on x.

In the second case, we know the function y: the integration is used to evaluate the area under the curve in a graph of y against x. This also allows us to determine the average value of y over some range of x.

4.2 Integrating a rate equation

In reaction kinetics we often have a *reaction rate equation* which expresses how some concentration c varies in time, i.e. the derivative dc/dt. An example is that of a first order rate law

$$dc/dt = -k\,c$$

where k is called the *rate constant*. We would integrate this equation to get an expression that tells us how c depends on time t. This new equation would have to be such that if we differentiate it with respect to t, we get $-kc$ back again.

A similar problem is that of a car moving with a particular *velocity*. Velocity is the rate of change of position s with time t, ds/dt. If we know the velocity, we can integrate to find the position s at any time t.

The simplest case arises if the speed is constant,

$$\frac{ds}{dt} = m$$

where m does not depend on s or t.

Exercise

Suppose that a car starts at $s = 0$ and is driven at a constant speed $m = 30$ mph.
How far will the car have travelled after (i) 30 mins, (ii) 1 hour, (iii) 2 hours?

(Answers: (i) 15 miles, (ii) 30 miles, (iii) 60 miles.)

In this case, we can show the distance travelled as a function of time by plotting a graph with a slope of 30 mph.

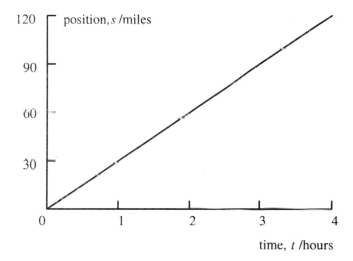

The position s is related to the time travelled by

$$s = 30\,\text{mph} \times t$$

i.e.

$$s = mt$$

where the speed m gives the slope of this graph (it is a straight line because the slope is constant).

We can check this result by differentiating the final equation with respect to t: from the rules given in the previous section,

Exercise

Differentiate $s = mt$ to find ds/dt

$$\frac{ds}{dt} =$$

We obtain $ds/dt = m$, which is the same as the equation we started with, so $s = mt$ is a correct *integrated form* of the original *differential equation*.

Exercise

For a zero-order chemical reaction the rate at which the concentration c of some product increases with time is given by

$$\frac{dc}{dt} = k$$

where k is the rate constant. Sketch the graph of c versus t and obtain an expression for c assuming that $c = 0$ initially.

(Answer: the graph is again a straight line emerging from the origin of slope k: the equation for c as a function of time is $c = kt$.)

If the initial concentration of the product had not been zero, but had been some value c_0, then the equation for c would be altered slightly

$$c = kt + c_0$$

with kt now giving how c increases above its initial value as time progresses.

If we differentiate this new equation, we still obtain $dc/dt = k$, because the constant term disappears on differentiating.

We can see the difference that c_0 makes by plotting a series of graphs

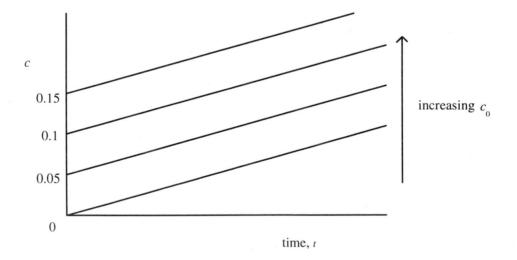

All the lines have the same gradient (slope $= k$) but they have different intercepts corresponding to their different initial concentrations. Because the intercept does not affect the slope and because it is a constant term that disappears under differentiation, all these lines fit the equation

$$dc/dt = k$$

This one differential equation thus fits a whole *family* of integrated forms, the members of the family differing by the exact value of the extra, constant term c_0.

4.3   Integration of polynomials

We examined a simple, special case in the previous section and saw that there is some relationship between the integrated and differential forms of an equation.

In general we are trying to find the equation for some quantity y in terms of another x if we know that dy/dx is a specific function g which may involve x or y (or even both)

$$dy/dx = g$$

For the moment we will concentrate on cases in which g only depends on x and is some polynomial, i.e. it involves terms like x^n.

If we take $y = x^{n+1}$ and differentiate it, we get $dy/dx = (n+1)\, x^n$.

There are thus two things that have happened to the right-hand side of the equation. First, the power of x has *decreased* by one, from $n+1$ to n. Second, a new factor of $(n+1)$ has appeared as a *multiplier*.

If integration is the inverse process, then it must reverse both these changes:

so we can expect that integration will (a) *increase* the power of x by 1 and (b) cause the term to be *divided* by some factor.

From these arguments, we can eventually conclude that the integral of x^n must be $x^{n+1}/(n+1)$. We write this as

$$\int x^n dx = \frac{1}{(n+1)} x^{n+1} + C$$

The left-hand side is read as 'the integral of x^n with respect to x'. Notice that just as we differentiate a given function with respect to some quantity, so we also integrate functions with respect to (the derivative of) some quantity.

The final term on the right-hand side is an *arbitrary constant*. We saw above, that integration produces a whole family of solutions differing only by an added (or subtracted) constant, so we need to write this possibility in explicitly.

We can follow the algebra behind the integration:

we have started with the information $\dfrac{dy}{dx} = x^n$

to find y, we effectively multiply the dx term up onto the right-hand side and we then add in the integral signs to give $\displaystyle\int dy = \int x^n dx$

the integral of dy is simply y $\displaystyle y = \int x^n dx$

finally, we evaluate the right-hand side using the rule given above to give

$$y = \frac{1}{(n+1)} x^{n+1} + C$$

Exercise

Check that this result does satisfy the original differential form of the equation by finding dy/dx

(Answer: $dy/dx = \dfrac{1}{(n+1)} \times \left[(n+1)x^n\right] + 0 = x^n$, as required.)

Exercises

Evaluate y in the following cases

(a) $\dfrac{dy}{dx} = x$ $y =$

(b) $\dfrac{dy}{dx} = x^4$ $y =$

(c) $\dfrac{dy}{dx} = 3x^5$ $y =$

(d) $\dfrac{dy}{dx} = x^{-2}$ $y =$

$$\int_{.} dy = \int x \, dx \ . \quad y + C_y = \frac{x^2}{2} + C_x$$
$$y = \frac{x^2}{2} + C_x - C_y = \frac{x^2}{2} + C$$

(Answers: (a) $y = \frac{1}{2}x^2 + C$, (b) $\frac{1}{5}x^5 + C$, (c) $\frac{3}{6}x^6 + C = \frac{1}{2}x^6 + C$, (d) $-\frac{1}{x} + C$.)

Exercise (d) indicates that this rule can be used even if the power n is negative, i.e. if $dy/dx = x^{-n} = 1/x^n$.

A special case arises if $n = 0$. This allows us to include the integration of a constant under this rule.

If we have the equation $dy/dx = k$, where k is a constant, we can re-write this as

$$dy/dx = k\, x^0$$

(because $x^0 = 1$). The rule above still works, so we get

$$\int k \, dx = kx + C$$

This result is consistent with the equation we wrote down by looking at the graph in the previous section.

There is, however, one other special case, which occurs if $n = -1$, i.e. if $dy/dx = 1/x$.

We cannot use the general formula above as the term $1/(n+1)$ then becomes $1/0$.
We can, however, refer back to the table of derivatives in the previous section to see which function gives $1/x$ when differentiated. Thus we find

$$\int \frac{1}{x} dx = \ln(x) + C$$

i.e. we obtain the natural logarithm of x (+ the arbitrary constant).

This result is particularly important in chemistry: it arises when dealing with radioactivity, first order kinetics and in thermodynamics.

Example

For a first order reaction, the rate at which the concentration c of the reactant changes is given by

$$\frac{dc}{dt} = -kc$$

If we re-arrange this and add the integral signs, we can obtain the equation

$$\int \frac{dc}{c} = -\int k \, dt$$

The right-hand side is simply the integration of a constant, which we have seen before, and gives $-kt + C_1$, where C_1 is the arbitrary constant.

The left-hand side has the form dx/x and so will integrate to give a logarithm term. We thus have

$$\ln(c) + C_2 = C_1 - kt$$

where C_2 is a second arbitrary constant.

In fact, we only need one arbitrary constant in these cases. The constant C_2 can be moved over to the right-hand side by the normal rules of algebra (subtracting it from each side). This gives a term $(C_1 - C_2)$ which is itself simply an unknown constant which we can denote as C. The result then is

$$\ln(c) = C - kt. \qquad \text{\textit{t is TIME.}}$$

We can also determine the value of the arbitrary constant if we have one more piece of information. Perhaps we know the initial concentration is c_0. This means that $c = c_0$ at $t = 0$. Substituting these values into the equation we obtain

$$\ln(c_0) = C$$

i.e. the arbitrary constant is equal to the logarithm of the initial concentration in this case, and

$$\ln(c) = \ln(c_0) - kt$$

If you have a set of numbers you can take their antilog and change their equation around.

ANTILOGS

Taking exponentials of both sides

$$c = c_0 e^{-kt}$$

$\left\{ \begin{array}{l} \ln 3 = \ln 5 - kt \\ 1.0986 = 1.609 - kt \end{array} \right.$

$3 = \dfrac{5}{e^{kt}}$

Exponential variations arise when the rate at which some quantity varies in time is linearly proportional to the value of that quantity.

Exercise

The rate of decay of a radioactive nuclide follows the equation

$$\frac{dN}{dt} = -\lambda N$$

1st order

P55

Find the expression for the dependence of N on t if $N = N_0$ at $t = 0$.

$dN = -\lambda N \, dt \quad \therefore \int \dfrac{dN}{N} = -\int \lambda \, dt \quad . \quad e^x + C_x = -\lambda t + C_y \quad . \quad e^x = -\lambda t + C$

$d\ln x = \dfrac{1}{x}$

(Answer: $\ln N = C - \lambda t$ where C is an arbitrary constant; $\ln N_0 = C$, so $\ln N = \ln N_0 - \lambda t$ and $N = N_0 e^{-\lambda t}$.)

VERY HANDY DERIVATION

produce a few exercises to go with it

Exercise

A car is driven with a constant acceleration, so that its velocity increases in time according to the equation

$$\frac{ds}{dt} = mt$$

where m is a constant. Find the position of the car after 10 minutes if $m = 5$ m min^{-2} assuming that $s = 0$ at $t = 0$.

(i) Begin by integrating the equation to find an equation for s in terms of m and t

(ii) Use the information about the initial position to determine the arbitrary constant

(iii) Substitute in the final value for t and the value of m to obtain s

(Answer: (i) $s = C + \frac{1}{2} mt^2$, (ii) $C = 0$; (iii) $s = \frac{1}{2} \times 10 \times 5^{2} = 125$ m.)

$= 250$

Exercises

(a) In the simplest form of a second order reaction, the concentration c of the reactant follows the rate equation

$$\frac{dc}{dt} = -kc^2$$

Integrate this to find an expression for c if the initial concentration is c_0.

(b) The variation of an entropy change ΔS with temperature is governed by the equation

$$d(\Delta S) = \Delta c_p \frac{dT}{T}$$

where Δc_p is the heat capacity change for the reaction.

Integrate this equation to find an expression for ΔS as a function of temperature assuming that Δc_p is independent of temperature.

If $\Delta S = 90$ J K^{-1} mol^{-1} at 298 K and $\Delta c_p = 10$ J K^{-1} mol^{-1}, calculate the value of the constant of integration C and then calculate ΔS at 350 K.

(Answers: (a) $\frac{1}{c} = \frac{1}{c_0} + kt$, so $c = \frac{c_0}{1 + kc_0 t}$ (see section 1.4 for this rearrangement), (b) $\Delta S = \Delta c_p \ln T + C$,

using the known value we find $C = 33$ J K^{-1}mol^{-1}; so $\Delta S = 10 \times \ln T + 33$ in general for this reaction; taking $T = 350$ K, we have $\Delta S = 91.6$ J K^{-1} mol^{-1}.)

In (b) above, we can find the constant of integration in a more general form.

We know the value of ΔS_{298K}: such data are frequently tabulated at this temperature. We can use this to express C from the integrated equation:

$$\Delta S_{298K} = \Delta c_p \ln(298K) + C \qquad \text{so} \qquad C = \Delta S_{298K} - \Delta c_p \ln(298K)$$

Inserting this into the general result, we then obtain for the entropy change ΔS_T at any temperature T

$$\Delta S_T = \Delta S_{298K} + \Delta c_p \ln(T / 298K)$$

In the last example, we may wish to proceed more accurately and allow for a temperature dependence of the heat capacity change. It is common to find heat capacity data expressed in a form such that Δc_p can be written as

$$\Delta c_p = a + bT + \frac{c}{T}$$

where a, b and c are given constants.

In this situation, the equation for $d(\Delta S)$ becomes

$$d(\Delta S) = \frac{a + bT + c/T}{T} dT$$

We can write this as

$$d(\Delta S) = \frac{a}{T} dT + b dT + \frac{c}{T^2} dT$$

Each term in the sum on the right-hand side can be integrated separately.

Exercise

Integrate the equation above to find the temperature dependence of ΔS

$$\Delta S = \qquad\qquad\qquad\qquad + C$$

(Answer: $\Delta S = a \ln T + bT - \dfrac{c}{T} + C$.)

4.4 Integrating other functions

Just as other functions such as e^x or $\sin x$ can be differentiated, so they can be integrated. A table of integrals (so called *indefinite integrals*) can be constructed analogous to the table in section 3.3.

This table gives $y = \int g(x)\mathrm{d}x$ for different functions $g(x)$.

$g(x)$	$y = \int g(x)\mathrm{d}x$
m	$mx + C$
x	$\frac{1}{2}x^2 + C$
x^n	$\frac{1}{n+1}x^{n+1} + C \quad (n \neq -1)$
x^{-n}	$-\frac{1}{(n-1)}x^{-(n-1)} + C$
$1/x$	$\ln(x) + C$
e^x	$e^x + C$
e^{ax}	$\frac{1}{a}e^{ax} + C$
$\sin(ax)$	$-\frac{1}{a}\cos(ax) + C$
$\cos(ax)$	$\frac{1}{a}\sin(ax)$

$n = -1$ (handwritten annotation to the left)

In each case, C is the arbitrary constant that arises under the integration

Exercises

Evaluate the following integrals

(a) $\int x^{1.5}\mathrm{d}x =$

(b) $\int x^{-1.5}\mathrm{d}x =$

(c) $\int e^{3x}\mathrm{d}x =$

(d) $\int \frac{1}{e^{3x}}\mathrm{d}x =$

(e) $\int \sin(\pi x)\mathrm{d}x =$

(Answers: (a) $\frac{2}{5}x^{2.5} + C$, (b) $-2x^{-1/2} + C$, (c) $\frac{1}{3}e^{3x} + C$, (d) $-\frac{1}{3}e^{-3x} + C$, (e) $-\frac{1}{\pi}\cos(\pi x) + C$.)

4.5   Integration between limits: definite integrals and areas under a curve

The discussion so far has been in the spirit of knowing the slope of a curve and using integration to find the function for the curve itself, i.e. finding y if we know dy/dx.

If we know the equation for y, then we can also use integration for another useful purpose: to find the area underneath the curve, as sketched in the figure below

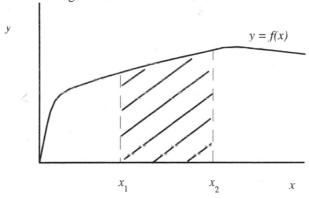

The area A shaded in the figure is obtained from the integral

$$A = \int_{x_1}^{x_2} y\,dx = \int_{x_1}^{x_2} f(x)\,dx$$

The suffices x_1 and x_2 on the integral signs are called the *lower* and *upper limits* and they indicate the range of x over which the integration is to be carried out.

Example

The application of this method can be illustrated by working out the area under the curve $y = 2x$ (i.e. the equation of a straight line of slope 2 emerging from the origin) between $x = 0$ and $x = 2$.

We proceed as follows:

(i) we write the integral to be evaluated with its upper and lower limits

$$A = \int_0^2 2x\,dx = 2\int_0^2 x\,dx$$

(note than any constant term multiplying the function can be brought out in front of the integral sign)

(ii) we evaluate the integral using the table given previously

$$A = \left[x^2 + C\right]_0^2$$

(note that the resulting expression is written inside square brackets with the upper and lower limits added as suffices)

(iii) the right-hand side is evaluated by substituting the upper limit into the resulting formula and then subtracting off the value of the expression evaluated with the lower limit.

In this case then we have

$$A = \left[2^2 + C\right] - \left[0^2 + C\right] = 4$$

We may note that the value of the arbitrary constant C did not have to be determined because it cancels when the upper and lower limit terms are subtracted. In fact, for such a *definite integral* (one with the upper and lower limits specified) we do not need to include the arbitrary constant in the integrated form.

We can check the result in this case using the graphical representation

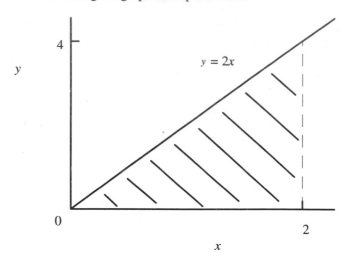

The integration evaluates the area of the shaded triangle. The area of a triangle is given by

$$Area = \tfrac{1}{2} \times base \times height$$

so in this case we have, with base = 2 and height = 4, $A = \tfrac{1}{2} \times 2 \times 4 = 4$ as determined by the integration process.

Exercises:

(a) Evaluate the area under the curve $y = 4x^2$ between $x = 0$ and $x = 5$

(b) Evaluate the area under the curve $y = e^{-3x}$ between $x = 0.1$ and $x = 0.2$

(c) Evaluate the area under the curve $y = e^{-3x}$ between $x = 0$ and $x = \infty$

(Answers: (a) $\left[\tfrac{4}{3}x^3\right]_0^5 = \tfrac{4}{3}\left[5^3 - 0^3\right] = \tfrac{500}{3}$, (b) $\left[-\tfrac{1}{3}e^{-3x}\right]_{0.1}^{0.2} = \left[-\tfrac{1}{3} \times (0.549 - 0.741)\right] = 0.064$,

(c) $\left[-\tfrac{1}{3}e^{-3x}\right]_0^{\infty} = \left[-\tfrac{1}{3} \times (0 - 1)\right] = \tfrac{1}{3}$ (this uses the results $e^{-\infty} = 0$ and $e^{-0} = 1$).)

The evaluation of the area under a curve arises frequently in thermodynamics. An example is the calculation of the work done during the compression or expansion of a gas.

Example

The work done in an expansion or compression in which the volume decreases by an infinitesimal amount dV against a pressure p, the work dw is given by

$$dw = -pdV$$

(for a compression the volume decreases and so dV is negative, the $-$ sign ensures that we follow the convention that the work done on the system is then a positive quantity: an expansion has $dV > 0$ and work is then done by the system).

The total work w done by a finite volume change from the initial volume V_{init} to a final volume V_{fin} is then obtained by integrating this expression

$$w = -\int_{V_{init}}^{V_{fin}} p\,dV = \int_{V_{fin}}^{V_{init}} p\,dV$$

The second form illustrates a rule:

Interchanging the upper and lower limits changes the sign of the integral

---

For *an expansion process* carried out on an ideal gas *at constant pressure*, this integration is relatively easy. The pressure can be brought out in front of the integral sign so we have simply

$$w = p\int_{V_{fin}}^{V_{init}} dV = p\left(V_{init} - V_{fin}\right) = -p\Delta V$$

where $\Delta V = V_{fin} - V_{init}$ is the *volume change* and will be negative for an expansion so w is positive : p is here the external pressure against which the expansion is 'working'.

For *a compression* carried out *at constant temperature*, the pressure in the system will increase. For an ideal gas, we have

$$pV = nRT, \text{ so } p = nRT/V.$$

We must substitute this into the equation so that all the terms 'under' the integral sign are constant or depend explicitly on V as this is the quantity we are integrating with respect to.

Thus we have

$$w = \int_{V_{fin}}^{V_{init}} \frac{nRT}{V}\,dV = nRT\int_{V_{fin}}^{V_{init}} \frac{dV}{V}$$

Exercise

Evaluate the integral in the above expression and hence determine the work done on the system if 0.1 mol of an ideal gas is compressed from 20 dm^3 to 15 dm^3 at 300 K

(Answer: $w = nRT \ln(V_{\text{init}} / V_{\text{fin}})$ — this uses the form $\ln(a)-\ln(b) = \ln(a/b)$;

$w = 0.1 \,\text{mol} \times 8.314 \,\text{J K mol}^{-1} \times 300 \text{K} \times \ln(20/15) = 71.8 \,\text{J}$.)

Another rule that arises when evaluating the area under a curve is:

A region that lies under the x-axis contributes a *negative* area to the total.

Example

The graph of $y = x^3$ has the appearance

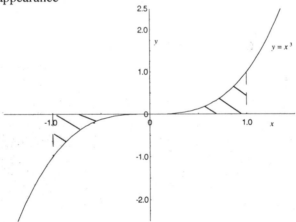

We can evaluate the area under the curve between $x = -1$ and $+1$ using the integration method

$$A = \int_{-1}^{1} x^3 dx$$

Exercise

Evaluate this definite integral and explain the result with reference to the graph above.

$$A =$$

(Answer: $A = \left[\frac{1}{4}x^4\right]_{-1}^{1} = \left[\frac{1}{4}\right] - \left[\frac{1}{4}\right] = 0$ as $1^4 = (-1)^4 = 1$. The two shaded areas in the graph are of equal area but opposite sign and so cancel exactly)

4.6 Some more definite integrals

There are a number of definite integrals whose values are important in chemistry. These have the general form

$$I = \int_0^\infty x^n e^{-ax^2}\,dx$$

where a is a constant and the power n may be 0, 1, 2 .....

The values for I are given below without proof and will be used in the next sections

n		I
0	$\int_0^\infty e^{-ax^2}\,dx$	$\frac{1}{2}\left(\frac{\pi}{a}\right)^{1/2}$
1	$\int_0^\infty x e^{-ax^2}\,dx$	$\frac{1}{2a}$
2	$\int_0^\infty x^2 e^{-ax^2}\,dx$	$\frac{1}{4}\left(\frac{\pi}{a^3}\right)^{1/2}$
3	$\int_0^\infty x^3 e^{-ax^2}\,dx$	$\frac{1}{2a^2}$
4	$\int_0^\infty x^4 e^{-ax^2}\,dx$	$\frac{3}{8}\left(\frac{\pi}{a^5}\right)^{1/2}$

4.7 Finding averages by integration

A third application of integration in chemistry is in evaluating the mean value of a quantity that changes continuously.

If we have discrete measurements of some quantity, perhaps the temperature of a liquid sampled at one minute intervals for 10 minutes

reading, i	1	2	3	4	5	6	7	8	9	10	11
time/min	0	1	2	3	4	5	6	7	8	9	10
T/K	303	301	302	301	303	304	305	306	304	302	301

then the average or mean value $\overline{T}$ is obtained by adding up the individual readings and dividing by the number of data points (here $n = 11$)

$$\overline{T} = \frac{\sum_{i=1}^{n} T_i}{n}$$

The subscript i here is known as the *index* of the quantity being added under the summation sign Σ and identifies a particular temperature in the sequence.

Exercise

Evaluate the mean temperature reading

$$\overline{T} \; =$$

(Answer: $\overline{T} = 3332\text{K}/11 = 302.9\text{K}$)

This approach is not applicable, however, if we have a quantity that is measured continuously as there are then effectively an infinite number of points to deal with.

Instead, we plot a graph of the quantity against, perhaps, time and then we use an adaptation of the method we practised in the previous section, i.e. we use the area under the curve.

Example

We may wish to know the average value during some period of time of the concentration c of a reactant undergoing a first order reaction.

The equation for the variation of the concentration with time is

$$c = c_0 e^{-kt}$$

What is the average value of c over the period $t = 0$ s to 10 s if c_0 is 0.5 M and $k = 0.1$ s^{-1}?

To find this we evaluate the area under the curve obtained by plotting c against time between $t = 0$ s and $t = 10$ s. This area can be obtained from the integral

$$A = \int_0^{10} c\,\mathrm{d}t = c_0 \int_0^{10} e^{-kt}\,\mathrm{d}t$$

The average value $\bar{c}$ is then obtained by dividing this integral by the time interval between the upper and lower limits, $\Delta t = 10$ s in this case.

$$\bar{c} = \frac{A}{\Delta t} = \frac{0.5 \int_0^{10} e^{-0.1t}\,\mathrm{d}t}{10}$$

Exercise

Evaluate the integral and hence determine the average concentration over the period specified.

(Answer: the integral is $A = -\dfrac{c_0}{k}\left[e^{-kt}\right]_0^{10}$, inserting the particular values we obtain

$A = -5\,\mathrm{M\,s} \times \left[e^{-1} - 1\right] = 3.16\,\mathrm{M\,s}$; then $\bar{c} = A / \Delta t = 3.16\,\mathrm{M\,s} / 10\,\mathrm{s} = 0.316\,\mathrm{M}$.)

Notice how the units eventually cancel to give a mean concentration in M as required. The time for the concentration of a reactant to fall to e^{-1} of its initial value is known as the *reaction lifetime*. The lifetime of a radioactive source (which also shows first order decay) is frequently quoted as being the time for the activity of the sample to fall to e^{-1} of its initial value.

Exercises

Evaluate the mean values of the following quantities
(a) of y if $y = x^2$, over the range $x = 1$ to 3

(b) of the pressure p during an isothermal expansion of 0.1 mol of ideal gas from $V = 15 \times 10^{-3}$ m^3 to 20×10^{-3} m^3 at 298 K (use $p = nRT/V$ and integrate with respect to V)

(Answers: (a) $\bar{y} = \frac{1}{3-1} \int_1^3 x^2 dx = \frac{1}{2} \left[\frac{1}{3} x^3 \right]_1^3 = \frac{26}{6}$; (b) $\bar{p} = \frac{1}{V_{\text{fin}} - V_{\text{init}}} \int_{V_{\text{init}}}^{V_{\text{fin}}} \frac{nRT}{V} dV = \frac{nRT}{(V_{\text{fin}} - V_{\text{init}})} \ln \left(\frac{V_{\text{fin}}}{V_{\text{init}}} \right)$,

substituting in the values, we obtain $\bar{p} = 14.26$ kPa .)

In general, the mean value $\bar{y}$ of some quantity y that is a function of x, over the range between an upper limit x_u and a lower limit x_1 is given by

$$\bar{y} = \frac{1}{(x_u - x_1)} \int_{x_1}^{x_u} y \, dx$$

with y being c and t being x in the previous example

This result can be interpreted graphically. The integral itself is the area under the curve, as described above. The term $(x_u - x_1)$ is the width of this area, and so $\bar{y}$ is actually the height of a rectangle of the same area as the area under the curve

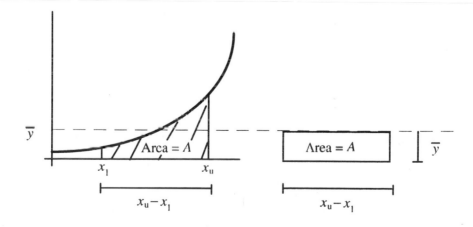

4.8   Mean values with probability distributions

In the kinetic theory of gases, an expression is derived for the *probability distribution* of the speed v.

This is an equation that allows us to calculate the probability that a molecule chosen at random will have a speed within a very narrow range close to a particular value.

The probability distribution function, usually denoted f in this instance is given by

$$f = 4\pi \left(\frac{m}{2\pi kT} \right)^{3/2} v^2 e^{-\frac{1}{2}mv^2/kT}$$

This looks less intimidating in the form

$$f = bv^2 e^{-av^2}$$

with $b = 4\pi(m/2\pi kT)^{3/2}$ and $a = m/2kT$.

The shape of this function is shown in the graph below and some important points, to be introduced below, are marked.

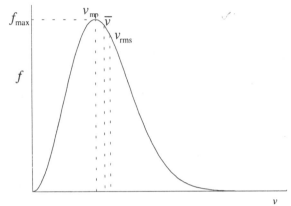

Probability functions generally have the property of being *normalised*.

Mathematically this means that if the function is integrated over all possible values of the quantity, the result is equal to one, i.e. there is a 100% probability that the observed result allows within the range of allowed values.

For the distribution above, this means

$$\int_0^\infty f \, dv = 1$$

as v can range from 0 to ∞.

Exercise

Use the integral forms given in section 4.6 to show that the probability distribution function above is normalised

(Answer: $b\int v^2 e^{-av^2} dv = \frac{b}{4}\left(\frac{\pi}{a^3}\right)^{1/2}$; substituting for a and b, this simplifies to 1.)

If the probability distribution function is known, then the average value of any quantity related to it can be determined by a suitable integration.

Example

If we wish to determine the average speed $\bar{v}$ we evaluate the integral

$$\bar{v} = \int_0^\infty (f \times v)\,dv$$

Substituting for f, this becomes

$$\bar{v} = \int_0^\infty b\,v^3 e^{-av^2}\,dv = \frac{b}{2a^2}$$

using the result from section 4.6. Substituting for a and b, we then have

$$\bar{v} = 4\pi \left(\frac{m}{2\pi kT}\right)^{3/2} \times \frac{1}{2}\left(\frac{2kT}{m}\right)^2 = \sqrt{\frac{8kT}{\pi m}}$$

for the *mean speed*

Exercise

Determine the mean value of v^2 by evaluating the integral $\int_0^\infty \left(f \times v^2\right)dv$. The square root of this quantity is known as the *root mean square speed* v_{rms}.

(Answer: $v_{rms}^2 = 3kT/m$ so $v_{rms} = \sqrt{3kT/m}$ compared with $\bar{v} = \sqrt{8kT/\pi m}$.)

Exercise

There is one more speed of interest and that is the most probable speed v_{mp}. This is simply the speed for which the probability distribution function f has its maximum value. Find this speed using the method of locating maxima presented in section 3.5

(Answer: $df/dv = 2bv\left(1 - av^2\right)e^{-av^2}$, v_{mp} makes this zero, so $v_{mp} = 1/\sqrt{a} = \sqrt{2kT/m}$.)

We can state the lessons of this section more formally:

the probability function $f\,dv$ allows us to find the fraction of molecules with a speed between the value v and $v + dv$, where dv is a small increase.

By integrating the function across a finite range, we obtain the fraction of molecules with speeds within that range:

the fraction of molecules with a speed v in the range $v_1 \le v \le v_2$ is given by $\int_{v_1}^{v_2} f\,dv$

The mean value of a function y involving v (e.g. v itself or v^2 etc.) is given by

$$\bar{y} = \int_0^\infty (f \times y)\,dv$$

4.9 Integration by substitution or by parts

There are a few 'tricks' involving 'clever' substitutions that are particularly useful to know about when faced with the problem of integration and one additional general method known as 'integration by parts'.

The 'tricks' are learnt mainly by practice. The main benefit of experience is that new integrals can often be recast in a form that fits one of the 'standard forms' listed earlier.

Of all the standard forms, perhaps the most important in chemistry is that leading to the logarithmic integral

$$\int \frac{1}{x} \, dx = \ln(x)$$

If we have a problem that involves polynomials in both the numerator and the denominator of a fraction, it is always worth looking to see if the numerator is in any way like the derivative of the denominator

---

Example

For

$$\int \frac{2x+3}{x^2+3x+2} \, dx$$

we can note that if $u = x^2 + 3x + 2,$

then $du/dx = 2x + 3.$

This latter result can be written as $du = (2x + 3)dx$

so we can rewrite the integral as $\int \frac{1}{u} \, du = \ln(u) = \ln(x^2 + 3x + 2)$

Exercises

Integrate the following equations

(a) $\int \frac{2}{1+2x} \, dx =$

(b) $\int \frac{1}{1+2x} \, dx =$

(c) $\int \frac{1}{1-2x} \, dx$

(Answers: (a) $\ln(1 + 2x)$, (b) $\frac{1}{2} \ln(1 + 2x)$, (c) $-\frac{1}{2} \ln(1 - 2x)$)

In case (a) the numerator is exactly the derivative of the denominator so we can substitute $u = 1 + 2x$ with the numerator then being du.

For cases (b) and (c) we need to introduce an extra factor to make the numerator equal to the derivative of the denominator: for (b) we can rewrite the integral as

$$\frac{1}{2} \int \frac{2}{1+2x} \, dx = \frac{1}{2} \int \frac{1}{u} \, du$$

for (c) we also need to introduce a change of sign, so we rewrite the integral as

$$-\frac{1}{2} \int \frac{-2}{1-2x} \, dx = -\frac{1}{2} \int \frac{1}{u} \, du$$

with $u = 1 - 2x$ in this case, so $du = -2dx$.

A method of some importance is that of *integration by parts*.

This is effectively the reverse of the use of the product rule for differentiation and is useful in many case where the function to be integrated can be written as the product of two simpler functions.

The rule for integration by parts is

$$\int u\,dv = uv - \int v\,du$$

So, if we can recognise our integral as involving a function u and the derivative of a second function v, we can recast the problem so it involves the integration of v with respect to u. This is worthwhile in some cases, as the latter may be recognisable as a standard form if we choose how we identify u and v carefully.

Example

The integral $\int x e^x\, dx$ can be evaluated using a combination of the substitution and integration by parts methods.

If we identify v as e^x, then $dv/dx = e^x$ or $dv = e^x dx$. If we also make the trivial identification $x = u$, we can rewrite the integral in the form of $u\,dv$

Noting that $du = dx$, this then becomes

$$x e^x - \int e^x\, dx = x e^x - e^x = (x-1)e^x$$

This approach is often useful with exponentials as they can be 'taken into the derivative' in this way.

Exercise

Use the method of integration by parts to evaluate the integral of $\ln(x)$.

For this it is helpful to make the following choices:

$$
\begin{array}{lllllll}
\text{let } u = \ln(x) & \text{so} & du/dx = & \text{or} & du = & dx \\
\text{let } v = x & \text{so} & dv/dx = & \text{or} & dv = & dx \\
\end{array}
$$

The integral can now be written in the form $\int u\,dv$ and solved by parts

(Answer: $\int u\,dv = uv - \int v\,du = x\ln(x) - \int x \times \dfrac{1}{x}\,dx = x\ln(x) - \int dx = x\ln(x) - x$)

Summary of Section

The material in this section has introduced the basic features of integration and three applications that arise in chemistry. The following concepts should be familiar:

- integration as the inverse of differentiation

- integration of simple powers of x, x^n

- integration of $1/x$ as a special case

- the constant of integration: why it occurs and how its value can be found

- integration of exponential, sine and cosine functions

- the upper and lower limits

- definite integrals

- the use of integration to find the area under a curve
$$A = \int_{x_1}^{x_2} y\,dx$$

- the use of integration to find average values of a function
$$\bar{y} = \frac{1}{(x_u - x_1)} \int_{x_1}^{x_u} y\,dx$$

- probability distributions
$$\bar{y} = \int_{x_1}^{x_u} f \times y\,dx$$

- integration by parts
$$\int u\,dv = uv - \int v\,du$$

$g(x)$	$y = \int g(x)dx$
m	$mx + C$
x	$\frac{1}{2}x^2 + C$
x^n	$\frac{1}{n+1}x^{n+1} + C$ $\quad (n \neq -1)$
x^{-n}	$-\frac{1}{(n-1)}x^{-(n-1)} + C$
$1/x$	$\ln(x) + C$
e^x	$e^x + C$
e^{ax}	$\frac{1}{a}e^{ax} + C$
$\sin(ax)$	$-\frac{1}{a}\cos(ax) + C$
$\cos(ax)$	$\frac{1}{a}\sin(ax)$

SECTION 5

Warming Down: Sines, Cosines and Complex Numbers

Sines, cosines and complex numbers

5.1 Circles and degrees

The area A and circumference L of a circle of radius r are given by

$$A = \pi r^2$$
$$L = 2\pi r$$

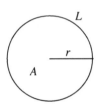

Exercises

(a) Calculate the area and circumference for a circle of radius 3 cm.

(b) What is the radius of a circle of area 19.6 m²?

(c) The diameter $d = 2r$: express the area and circumference in terms of d

(d) What is the ratio of the area to the circumference for a circle of diameter 0.1 m?

(Answers: (a) $A = 28.3$ cm², $L = 18.85$ cm; (b) 2.5 m; (c) $A = \frac{1}{4}\pi d^2$, $L = \pi d$;
(d) $A/L = \frac{1}{2}r = \frac{1}{4}d = 0.025$ m .)

A full circle corresponds to 360°. The angle in a semicircle is half that, i.e. 180° and a right angle (one quarter of a circle) corresponds to 90°

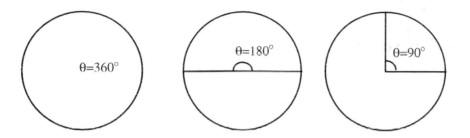

Exercise

In the circle, sketch the angles corresponding to (a) 60°, (b) 120° and (c) 270°

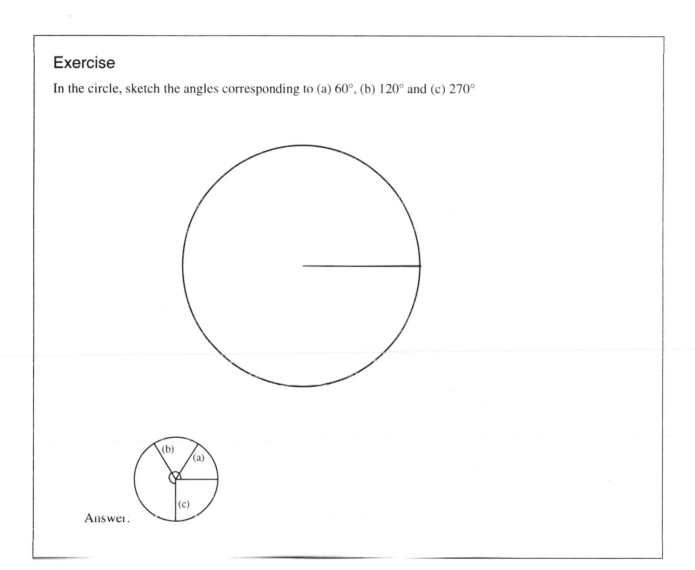

Answer.

5.2   Radians

There is an alternative method of describing angles involving the units of *radians*.

> The radian system expresses an angle in terms of the ratio of the radius of a circle and the length *s* of the *arc* that *spans* the angle.

This is most easily seen in terms of the following picture

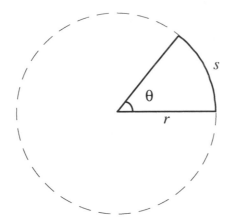

The angle is obtained by dividing *s* by *r*

$$\theta = \frac{s}{r} \text{ radians}$$

For a full circle, the arc is simply equal to the circumference $s = L$. In that case, then

$$\theta = \frac{L}{r} = \frac{2\pi r}{r} = 2\pi \text{ radians}$$

This allows us to obtain a method for converting between angles in degrees and radians. The angle for a full circle is 360°, so we have

$$360° = 2\pi \; radians$$

Exercise

Express the following angles in radians

(a) 180° =

(b) 90° =

(c) 60° =

(Answers: (a) π radians, (b) $\pi/2$ radians, (c) $\pi/3$ radians.)

In general, the conversion can be made by the formula

$$n° = \frac{360°}{2\pi} \times n \text{ radians} \qquad \text{or} \qquad n \text{ radians} = n° \times \frac{2\pi}{360°}$$

Exercise

Use this equation to complete the following table

angle	angle/radians
0°	
30°	
	π/4
60°	
90°	
	2π/3
180°	
	3π/2
360°	

The number π plays an important role in measurements in radians, just as e plays an important role as the base for natural logarithms and exponentials.

As chemists, we normally have to work in terms of radians and natural logs rather than degrees and logs to the base 10. Most calculators can be switched to 'radian mode' and almost all computers work in these 'natural units'.

Exercise

Use the radian mode on your calculator to evaluate the following:

(a) $\sin(1) =$

(b) $\sin(\pi/2) =$

(c) $\cos(1.047) =$

(Answers: (a) 0.84, (b) 1, (c) 0.5)

5.3   Triangles

The right-angled triangle shown below has three sides known as the *opposite O, adjacent A* and *hypotenuse H* relative to the angle θ

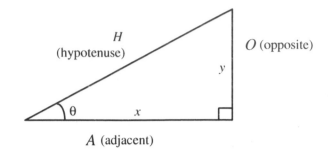

If the length of the adjacent is x and that of the opposite is y, then the length h of the hypotenuse can be calculated from *Pythagoras's theorem*

$$h^2 = x^2 + y^2 \qquad \text{or} \qquad h = \sqrt{x^2 + y^2}$$

The third angle, φ, can also be found by noting that the sum of the three angles of any triangle must add up to 180°, so

$$\theta + \phi + 90° = 180°$$

There are some important *trigonometric functions* associated with such triangles.

The *sine* of an angle, sin(θ) is the ratio of the length of the opposite to the length of the hypotenuse

$$\sin(\theta) = \frac{O}{H} = \frac{y}{\sqrt{x^2 + y^2}}$$

The *cosine* of an angle, cos(θ) is the ratio of the length of the adjacent to the length of the hypotenuse

$$\cos(\theta) = \frac{A}{H} = \frac{x}{\sqrt{x^2 + y^2}}$$

The *tangent* of an angle, tan(θ) is the ratio of the length of the opposite to the length of the adjacent

$$\tan(\theta) = \frac{O}{A} = \frac{y}{x}$$

We may also note that

$$\tan(\theta) = \sin(\theta)/\cos(\theta)$$

and that the tangent is also the gradient of the hypotenuse

Exercise

A triangle has an opposite of length $y = 3$ and an adjacent of length $x = 4$.

(a) sketch the triangle approximately to scale

(b) calculate the length of the hypotenuse

$$h \; =$$

(c) calculate the values of the functions

$$\sin(\theta) \; =$$

$$\cos(\theta) \; =$$

$$\tan(\theta) \; =$$

(Answers: (b) $h = \sqrt{3^2 + 4^2} = 5$, (c) $\sin(\theta) = 0.6$, $\cos(\theta) = 0.8$, $\tan(\theta) = 0.75$.)

From the definitions above we also have

$$\sin^2(\theta) + \cos^2(\theta) = 1$$

Here the notation $\sin^2(\theta)$ is used to denote the square of the sine function, i.e.

$$\sin^2(\theta) \; = \; \sin(\theta) \times \sin(\theta).$$

This is actually just a restatement of Pythagoras's theorem: we can check it using the results of the exercise above.

Exercises

(a) calculate the length of the hypotenuse for a triangle with an opposite of length $y = 1$ if $\sin(\theta) = 0.5$,

$$h \; =$$

(b) from y and h, calculate the length x of the adjacent

$$x \; =$$

(c) what are the values for

$$\cos(\theta)$$

and $\tan(\theta)$

(Answers: (a) $h = 2$, (b) $x = \sqrt{h^2 - y^2} = \sqrt{3}$, (c) $\cos(\theta) = \sqrt{3}/2 = 0.866$, $\tan(\theta) = 1/\sqrt{3} = 0.577$: the angle itself here is $\theta = 30°$ or $\pi/6$ radians.)

5.4   Trigonometric functions

The trigonometric functions introduced in the previous section are very important in chemistry. They are encountered in the *particle in a box* problem in elementary courses on quantum mechanics and arise whenever we have to deal with forces acting at angles etc. It is worthwhile examining some of their basic properties and learning to recognise some 'special' values.

The values of $\sin(\theta)$, $\cos(\theta)$ and $\tan(\theta)$ depend on the value of the angle θ itself.

We can find the values of these functions for any angle — positive or negative, no matter how large or small.

The variation of $\sin(\theta)$ with θ is shown below

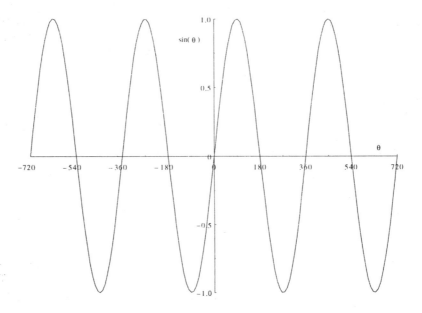

sin(θ) is a *periodic* function:

      the maximum value of $\sin(\theta)$ is $+1$
      the minimum value of $\sin(\theta)$ is -1

      sin(θ) repeatedly passes through 0 as θ varies

An additional feature from the shape of the graph is that

$$\sin(\theta) = -\sin(-\theta)$$

The zeros of the function are particularly important. We may note that:

$$\sin(\theta) = 0 \text{ if } \theta = -360°, -180°, 0, 180°, 360°, \text{ etc.}$$
or $$\sin(\theta) = 0 \text{ if } \theta = -2\pi, -\pi, 0, \pi, 2\pi \text{ etc. radian}$$

A general statement of this result is that

$$\sin(\theta) = 0 \qquad \text{if} \qquad \theta = \pm n\pi \text{ radians or } \pm n \times 180°$$

where n is any integer or zero

Exercise

Try to write a sentence 'translating' the above equation into words.

(A possible answer: The sine of an angle θ is equal to zero if θ is equal to zero or to π multiplied by an integer.)

Also, we see that

$$\sin(\theta) = +1, \qquad \text{if} \qquad \theta = -270°, +90° \text{ or } -3\pi/2, +\pi/2 \text{ radians etc.}$$

We can write this in a general form

$$\sin(\theta) = +1, \qquad \text{if} \qquad \theta = \frac{1}{2}(2n+1)\pi \text{ radians}$$

where n is zero or an *even* positive or negative integer.

$\sin(\theta)$ has the value -1 when $\theta = -90°, +270°$ or $-\pi/2, +3\pi/2$ radians etc.

In general

$$\sin(\theta) = -1, \qquad \text{if} \qquad \theta = \frac{1}{2}(2n+1)\pi \text{ radians}$$

where n is an *odd* positive or negative integer.

Exercises

Quote all answers in radians
    (a) Find *three* angles, in addition to those given above, for which $\sin(\theta) = 0$

    (b) Find *two* angles, in addition to those given above, for which $\sin(\theta) = +1$

    (c) Find *two* angles, in addition to those given above, for which $\sin(\theta) = -1$

(Possible answers: (a) 3π, -3π, 4π; (b) $5\pi/2$, $9\pi/2$, (c) $7\pi/2$, $-5\pi/2$.)

These results can be checked using the sine function on a calculator in radian mode.

Exercise

Using the radian mode on a calculator, evaluate the sin(θ) for the following

(a) $\theta = 0.1$ $\sin(\theta) =$

(b) $\theta = 0.01$ $\sin(\theta) =$

(c) $\theta = 0.001$ $\sin(\theta) =$

(d) $\theta = -0.001$ $\sin(\theta) =$

The answers for this exercise should illustrate one final property of the sine function for angles expressed in radian mode

$$\sin(\theta) \approx \theta \qquad \text{if} \qquad \theta \text{ is small}$$

The variation of $\cos(\theta)$ with θ is shown below

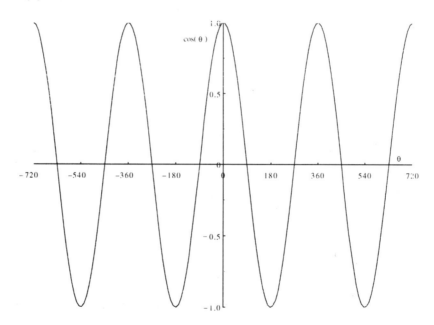

Again, this is a periodic function that oscillates between a minimum of -1 and a maximum value of $+1$.

The graph is symmetric about the y-axis, so

$$\cos(\theta) = \cos(-\theta)$$

The zeros of $\cos(\theta)$ occur whenever the angle is a right angle

$$\cos(\theta) = 0 \qquad \text{for} \qquad \theta = \pm90°, \pm270° \text{ or } \pm\pi/2, \pm3\pi/2 \text{ radians etc.}$$

Written most generally, this is

$$\cos(\theta) = 0 \qquad \text{for} \qquad \theta = \pm\frac{1}{2}(2n+1)\pi \text{ radians}$$

We may also note that:

$$\cos(\theta) = +1 \quad \text{for} \quad \theta = 0, 2\pi, 4\pi, \text{ i.e. for } 2n\pi$$

$$\cos(\theta) = -1 \quad \text{for} \quad \theta = \pi, 3\pi, 5\pi, \text{ i.e. for } (2n+1)\pi$$

where *n* is any integer.

If the angle θ is measured in radians, then for small θ

$$\cos(\theta) \approx 1 - \theta$$

The variation of $\tan(\theta)$ with θ is slightly more complex.

The tangent is not restricted to values between +1 and −1, but can be infinitely large (positive or negative).

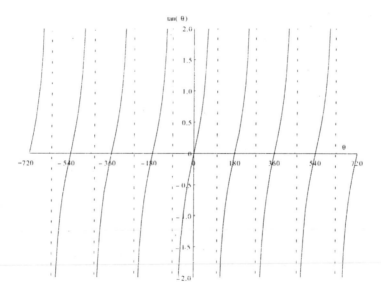

The infinities arise whenever the angle is a right angle as the gradient is then infinitely steep.

$\tan(\theta)$ is zero at the angles for which $\sin(\theta)$ is zero, i.e. at $0°$, $\pm 180°$, $\pm 360°$ etc.

$$\tan(\theta) = 0 \quad \text{for} \quad \theta = \pm n\pi$$

where *n* is any integer.

$$\text{For small angles,} \quad \tan(\theta) \approx \theta \qquad (\theta \text{ in radians})$$

Exercise

A '45°-triangle' has $\theta = 45°$ and is such that the lengths of the adjacent and opposite are equal, $x = y$.

(a) Sketch this triangle approximately to scale

(b) If $x = 1$, calculate the length of the hypotenuse using Pythagoras's theorem

$$h =$$

(c) Calculate the sine, cosine and tangent of θ (without the help of a calculator)

$$\sin(\theta) =$$

$$\cos(\theta) =$$

$$\tan(\theta) =$$

(d) What is the value of the third angle ϕ: quote this and θ in radians

(Answers: (b) $h = \sqrt{2}$, (c) $\sin(\theta) = \cos(\theta) = 1/\sqrt{2} = 0.707$, $\tan(\theta) = 1$, (d) $\phi = 90° - \theta = 45° = \pi/4$ radians.)

The radian mode is particularly important when performing integration of trigonometric functions or over circular regions.

Example

The simplest wavefunction ψ that arises in the 'particle on a ring' model for rotation and which corresponds to the particle having zero rotational energy is

$$\psi = A$$

where A is a constant that has to be determined.

The square of the wavefunction ψ^2 is actually a probability distribution function (see section 4.8) and so it must satisfy the criterion of being normalised — in other words as the particle must be somewhere on the ring, if we take $\psi^2 = A^2$ and integrate it over the whole ring, we must get the value of 1.

The required integral is $\int_0^{2\pi} A^2 d\theta = A^2 \int_0^{2\pi} d\theta$

because the angle θ varies from 0 to 2π as we go round the ring.

Exercise

Evaluate the integral and then, using the condition that this must be equal to 1, find the value of the constant A.

(Remember that $\int dx$ is simply equal to x)

(Answer: $A^2 \int_{0}^{2\pi} d\theta = A^2 [\theta]_0^{2\pi} = 2\pi A^2$: for $2\pi A^2 = 1$, we need $A = \sqrt{1/2\pi}$.)

5.5   Some special cases

Exercise

Complete the following table. This lists some important 'special' results for the trigonometric functions which arise frequently in chemistry

θ	θ/radian	$\sin(\theta)$	$\cos(\theta)$	$\tan(\theta)$
$0°$	0	0		
$30°$				
$45°$				
$60°$				
$90°$	$\pi/2$		0	∞
$120°$				
$180°$	π			
$270°$	$3\pi/2$			$-\infty$
$360°$	2π			
		$\sin(\theta) = -\sin(-\theta)$ $\sin(\theta)= 0$ for $\theta = \pm n\pi$	$\cos(\theta) = \cos(-\theta)$ $\cos(\theta) = 0$ for $\theta = \pm(2n+1)\pi$	

5.6   Inverse trigonometric functions

The previous sections have provided practice in answering the question 'What is the sine, cosine or tangent of a known angle θ'?

We now turn to the reverse process, i.e. we will be concerned with the question 'What angle has a particular value for its sine, cosine or tangent'?

Example

What is the angle θ for which $\sin(\theta) = 0.5$?

From the above table, we can see that $\sin(\theta) = 0.5$ when $\theta = 30°$ or $\pi/6$ radians

---

Two different notations can be found in books for the function that is the inverse of the sine function:

either $\qquad \theta = \sin^{-1}(x)$

or $\qquad \theta = \text{arc } \sin(x)$

These both specify the angle θ whose sine is equal to the value x.

The first form, $\sin^{-1}(x)$, is common on calculators, but runs the risk of being confused with the value of $1/\sin(\theta)$. (Remember $\sin^2(\theta)$ is used to denote $\sin(\theta) \times \sin(\theta)$: if we need to denote $1/\sin(\theta)$ we should use $(\sin(\theta))^{-1}$.)

The arc $\sin(x)$ notation does not run the risk of this confusion, so will be used here.

As $\sin(\theta)$ and $\cos(\theta)$ lie in the range from -1 to $+1$, we can only evaluate the arc $\sin(x)$ and arc $\cos(x)$ functions for x between -1 and $+1$

---

there are no angles that have a sine or cosine greater than $+1$ or less than -1.

Otherwise, we can simply use the appropriate function buttons on a calculator or recognise a special case to find the angle θ from the arc sin or arc cos.

Example

Evaluate (i) arc $\sin(0.866)$, (ii) arc $\cos(0.707)$ and (iii) arc $\tan(1)$:

(i) $\qquad \theta = 60° = 1.047 \, (= \pi/3)$ radians

(ii) $\qquad \theta = 45° = 0.785 \, (= \pi/4)$ radians

(iii) $\qquad \theta = 45° = 0.785 \, (= \pi/4)$ radians

Exercises

Evaluate the following angles

(a) arc sin(−0.4) =

(b) arc tan(5) =

(c) arc cos(0.95) =

(Answers: (a) $\theta = -23.6° = -0.41$ radians, (b) $\theta = 1.37$ radians, (c) $\theta = 0.318$ radians.)

It is, however, important to remember that the angle given as the answer by your calculator is not the only possible answer.

For instance, the sine of **all** of the following angles is 0.5, so all have the same arc sin

$$\theta = 30°, 390°, -330°, 750° \quad \text{or} \quad \pi/6, 13\pi/6, -11\pi/6, 25\pi/6 \text{ radians}$$

Calculators will normally only return the corresponding angle that lies in the range −180° to +180° (or −π to +π in radian mode).

Exercise

The position of certain bands in the infra-red spectrum of SO_2 can be used to determine the angle θ of the O–S–O bonds. The analysis leads to the equation

$$\sin^2\left(\tfrac{1}{2}\theta\right) = 0.769$$

Use this to find θ, which is known to lie in the range $90° < \theta < 180°$

(Answer: take the square root to find $\sin\left(\tfrac{1}{2}\theta\right) = 0.877$, so $\theta = 2$ arc sin(0.877). The calculator returns $\theta = 61.3°$, giving $\theta = 122.5°$, which lies in the correct range. We might also note that the angle 118.7° has a sign of 0.877 so could have been accepted, but the corresponding result for $\theta = 237.4$ lies outside the range known in advance: in fact this gives the angle $360° - \theta$.)

5.7   Imaginary numbers

In Section 2, we considered how to obtain the square roots of any positive number. It was also stated that negative numbers do not have (real) square roots, i.e. there is no (real) number that produces a negative result when multiplied by itself.

The latter point arises because when any two negative numbers are multiplied together, the − signs cancel leaving a positive quantity.

We now will see how some progress can be made for negative numbers and, in the process, see what is meant by the phrase 'real' written in brackets above.

We need first to recall a few simple rules:

---

any positive number can be written as itself multiplied by +1

　　　i.e. $5 = 1 \times 5$ or $x = 1 \times x$ in general

any negative number can be written as a positive number multiplied by −1

　　　i.e. $-5 = -1 \times 5$ or $-x = -1 \times x$ in general

the square root of two numbers multiplied together is equal to the product of the square roots of the two numbers

$$\sqrt{a \times b} = \sqrt{a} \times \sqrt{b}$$

We can use the last two of these to take one step forward in evaluating the square root of a negative number:

(i) we rewrite the negative number as a positive number $\times -1$

(ii) we take the square roots separately

Example

If $x = -9$, then we proceed as above

(i)　　　$x = -1 \times 9$

(ii)　　　$\sqrt{x} = \sqrt{-1} \times \sqrt{9} = 3 \times \sqrt{-1}$

In fact, there is another root given by $-3 \times \sqrt{-1}$ as $\sqrt{9} = \pm 3$

In general then, we can write

$$\sqrt{-a} = \sqrt{a} \times \sqrt{-1}$$

---

and $\sqrt{a}$ can be evaluated by the methods of section 2.2.

Exercise

Use the above method to rewrite the following square root problems

(a) $\sqrt{-25} =$

(b) $\sqrt{-40} =$

(Answers: (a) $\pm 5 \times \sqrt{-1}$, (b) $\sqrt{40} \times \sqrt{-1} = \pm 6.32 \times \sqrt{-1}$.)

We have seen above that the problem of finding the square root of any negative number can be *transformed* into the problem of finding the square root of a positive number multiplied by the square root of -1.

This is no great advance if we cannot find the number that is the square root of -1.

We know that there is no real number which gives -1 when squared, but perhaps we could, for now, suspend our disbelief and *imagine* that such a number does exist. We can give this number a symbol, and it is conventional to use i

$$i = \sqrt{-1} \qquad \text{so} \qquad i^2 = -1$$

We can now express any square root of a negative number in the form

$$\sqrt{-a} = \sqrt{a} \times \sqrt{-1} = \sqrt{a} \times i$$

Example

Find the square roots of -25

We use the above formula with $a = 25$, so $\sqrt{25} = \pm 5$, giving

$$\sqrt{-25} = \pm 5i$$

(Note the convention that the $\times$ sign is not written explicitly.)

Exercises

Express the following square roots in terms of i
(a) $\sqrt{-9} =$

(b) $\sqrt{-40} =$

(Answers: (a) $\pm 3i$, (b) $\pm 6.32i$.)

5.8   Complex numbers and arithmetic

We now have two types of number to deal with: real numbers, with which we are fairly familiar and, now, *imaginary* numbers that involve i the square root of -1.

In some situations we have to combine these two sets to produce a *complex number* which has both real and imaginary parts.

The general form of a complex number c is

$$c = a + b\text{i}$$

An example, with $a = 1$ and $b = 2$ is $c = 1 + 2\text{i}$.

A complex number $c = a + b\text{i}$ consists of a *real part a* and an *imaginary part b*i.

Complex numbers can be added or subtracted fairly simply. We just treat the real and imaginary parts separately and use the usual rules of arithmetic.

Example

If $c = 1 + 2\text{i}$ and $d = 2 - 3\text{i}$, then

$$c + d = (1 + 2) + (2 - 3)\text{i} = 3 - \text{i}$$

In general, if $c = a + b\text{i}$ and $d = e + f\text{i}$,

$$c + d = (a + e) + (b + f)\text{i}$$
$$c - d = (a - e) + (b - f)\text{i}$$

Exercises

Evaluate the following

(a) $(14 + 6\text{i}) + (6 + 3\text{i}) =$

(b) $(7 + 2\text{i}) - (4 - 3\text{i}) =$

(Answers: (a) $20 + 9\text{i}$, (b) $3 + 5\text{i}$.)

Complex numbers can also be multiplied using the regular rules of algebra developed in Section 1.4.

Example

Multiply $c = 1 + 2i$ by $d = 3 + 4i$

$$(1 + 2i) \times (3 + 4i) \qquad = 3 \times (1 + 2i) + 4i \times (1 + 2i)$$

$$= 3 + 6i + 4i + 8i^2$$

$$= 3 + 10i + 8i^2$$

To complete this, we notice that the final term involves i^2. We can simplify this if we remember that $i^2 = -1$. The final equation can then be rewritten as

$$(1 + 2i) \times (3 + 4i) = 3 + 10i - 8$$

i.e. $$= -5 + 10i$$

The result of multiplying two complex numbers is another complex number

---

Exercises

Multiply the following complex numbers

    (a) $(4 + 5i) \times (3 + 2i) =$

    (b) $(3 - 2i) \times (1 + 2i) =$

(Answers: (a) $2 + 23i$, (b) $7 + 4i$)

An important quantity is the *complex conjugate* of any complex number.

If $\qquad\qquad\qquad\qquad c = a + bi,$
then its complex conjugate $\quad c^* = a - bi$

---

is obtained by changing the sign of the imaginary part.

Exercise

If $c = 1 + 2i$, write down the complex conjugate c^*.

$$c^* =$$

Multiply c by c^* using the method above.

$$c^* \times c =$$

(Answers: $c^* = 1 - 2i$, $c^* \times c = 5$.)

If a complex number is multiplied by its conjugate, the answer is simply a real number.

In general, if $c = a + bi$, the product of c and c^*, which is usually written as c^*c is

$$c^*c \ = \ a^2 \ + \ b^2$$

The *modulus* of a complex number c is obtained by taking the square root of c^*c

$$|c| = \sqrt{c^*c} = \sqrt{a^2 + b^2}$$

This gives us a measure of the size or *magnitude* of a given complex number and allows us to compare two different complex numbers.

The complex conjugate can be used to perform the division of one complex number by another.

Example

To find the expression for $(1 + 2i) \div (1 + 3i)$, we proceed as follows:

(i) we multiply top and bottom by the conjugate of the denominator, here $(1 - 3i)$

$$\frac{(1+2i)}{(1+3i)} = \frac{(1+2i)\times(1-3i)}{(1+3i)\times(1-3i)}$$

(ii) multiplying out the terms in the numerator and denominator, we obtain

$$\frac{(1+2i)\times(1-3i)}{(1+3i)\times(1-3i)} = \frac{1+2i-3i-6i^2}{1+3i-3i-9i^2} = \frac{7-i}{10} = 0.7 - 0.1i$$

The multiplying out leaves just a real number in the denominator which we can divide into the complex number in the numerator.

Exercise

Evaluate the following

$$\frac{(2+3i)}{(1+4i)} =$$

(Answer: $(14 - 5i)/17$.)

Complex numbers sometimes arise involved with the exponential function, i.e. we have to use terms involving e^{iax}

This occurs as this is a convenient alternative way of representing complex functions that involve sin and cos terms and uses the results that

$$e^{iax} = \cos(ax) + i\sin(ax)$$

If we differentiate e^{iax} twice, we obtain

$$d^2(e^{iax})/dx^2 = i^2 a^2(e^{ix}) = -a^2(e^{ix})$$

compared with

$$d^2(e^{ax})/dx^2 = a^2(e^{ax}) = +a^2(e^{ax})$$

with the i in the exponent leading to the emergence of a $-$ sign in the second derivative. This is useful in some quantum mechanics problems such as the particle-in-a-box model.

Summary of Section

The material in this section has revised the basic trigonometric functions sin, cos and tan and introduced the idea of radian measure. It has also dealt with imaginary and complex numbers that arise when we take the square root of negative numbers. The following concepts should be familiar:

- the relationship between radius, area and circumference of a circle $A = \pi r^2, \quad L = 2\pi r$

- the measurement of angle by degrees

- the conversion between degrees and radians $\theta^o = \dfrac{\theta}{2\pi} \times 360^o$

- the basic features of a right angle triangle including the opposite, adjacent and hypotenuse

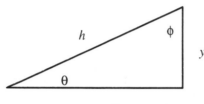

- $\quad x \qquad\qquad\qquad\qquad h^2 = x^2 + y^2; \; \theta + \phi = 90^o$

- the sine, cosine and tangent functions and their relationship to the sides of a triangle

- $\qquad\qquad\qquad\qquad \sin\theta = y/h, \; \cos\theta = x/h, \tan\theta = y/x$

- the relationship between $\sin\theta$, $\cos\theta$ and tan θ and Pythagoras theorem $\sin^2\theta + \cos^2\theta = 1$

- the variation of $\sin\theta$, $\cos\theta$ and $\tan\theta$ with the angle θ and, in particular, the angles for which $\sin\theta$ etc. $= 0$ or 1

- the meaning of the inverse functions arc sin, arc cos and arc tan

- imaginary numbers as square roots of negative numbers $\sqrt{-a^2} = \pm ai$

- complex numbers, their complex conjugates and magnitude $c = a + bi, \; c^* = a - bi, \; c^*c = a^2 + b^2$

- addition, subtraction, multiplication and division of complex numbers

$$(a + bi) + (c + di) = (a + c) + (b + d)i$$
$$(a + bi) - (c + di) = (a - c) + (b - d)i$$
$$(a + bi) \times (c + di) = (ac - bd) + (ad + bc)i$$
$$\frac{(a + bi)}{(c + di)} = \frac{(a + bi)}{(c + di)} \times \frac{(c - di)}{(c - di)} = \frac{(ac + bd) + (bc - ad)i}{c^2 + d^2}$$

SECTION 6

Relaxing: Statistics – Means and Deviations

Statistics: Means and Deviations

6.1 Repeated measurements: distributions

If some quantity is sampled by a series of experimental measurements, it is almost inevitable that there will be (at least) minor variations in the value observed from reading to reading.

Example

The concentration in parts per million (ppm) of Cu^{2+} ions as a pollutant in water is measured in seven samples collected from the same source, giving

$[Cu^{2+}]$/ppm	17	24	19	12	21	17	16

It is important to be able to summarise these observations, in a concise form, that indicates both a representative value of the set and the spread of the observations about this representative value.

For this we use the *mean* and the *variance* and *standard deviation*. (Other 'averages' and distribution measures are used less frequently in chemistry).

6.2 Mean value

The mean value of a sequence of readings is calculated simply by adding up their values and dividing this total by the number of readings.

This instruction is written mathematically as

$$\bar{x} = \frac{\sum_{i=1}^{n} x_i}{n}$$

Here $\bar{x}$ is the mean value of x, x_i is the value of an individual reading with the subscript i being the *index* which is simply a label indicating perhaps the order in which the readings were taken. The total number of readings is n

Exercise

Evaluate the average value of the Cu^{2+} concentration from the above set.

(Answer: Here $n = 7$ and $\sum_{i=1}^{n} [Cu^{2+}] = 126\,ppm$ so $\overline{[Cu^{2+}]} = 18\,ppm$)

Mean values can often be found directly using special function keys or the statistical mode on modern calculators.

6.3 Distributions

The mean may be a helpful summary of a data set, but cannot convey everything we wish to know.

For instance, the following two sets

set 1	$x_i =$	0	50	100
set 2	$x_i =$	49.9	50.0	50.1

have the same mean value $\overline{x} = 50.0$, but simply reporting this hides an important story about the way in which the individual readings are distributed about the average.

(We could argue that the mean for the first data set should be reported with less precision as 50 not 50.0, but even this refinement does not add much information.)

A very simple way of indicating the *range* around the mean covered by the observed data would be to include information about the highest and lowest values. We could do this by reporting the mean as

$$\overline{x} = 50\,(\pm 50) \qquad \text{for set 1}$$
$$\overline{x} = 50.0\,(\pm 0.1) \qquad \text{for set 2}$$

The $\pm$ term conveys additional information. However, if we now make another measurement and if this were, say, 49.7 we would have to change the range in set 2.

The most common quantities used to convey the distribution of values about the mean are the *variance* σ^2 and the *standard deviation* σ.

To understand the role played by these quantities, we need to discuss the concept of a *normal distribution*.

The normal distribution is the expected distribution of a set of data values about the 'real value' arising from presence of random experimental or sampling errors.

To observe the full distribution curve, we would need to collect an infinite number of data points. If we then plot the number of times a particular value of the quantity is found we get a curve of the form shown below

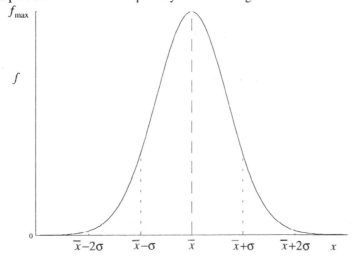

Such a curve is known as a *frequency* or *probability distribution curve*. For the normal distribution, the maximum occurs at the mean $\bar{x}$. The formula for the curve is

$$f = \frac{1}{\sqrt{2\pi}} \times \frac{1}{\sigma} e^{-(x-\bar{x})^2/2\sigma^2}$$

The first term $1/(\sqrt{2\pi}\,\sigma)$ is a normalising factor ensuring that the total area under the curve is equal to 1. The variance σ^2 arises in the exponent to scale the departure of x from its mean value $\bar{x}$.

From the graph, we can see that the probability of observing values of x either 'much larger' or 'much smaller' than $\bar{x}$ is low. In fact, we can be specific about what is meant by 'much larger' or 'much smaller'. We mean that the difference $x - \bar{x}$ becomes large compared with the standard deviation σ.

By integrating the normal distribution curve between the upper and lower limits of a particular range we can calculate the area under the curve in this range. This immediately gives the probability of observing values of x within the range. The integral is not particularly nice, but is related to the *error function* which has been calculated and tabulated on our behalf by earlier workers. The important results are:

68% of the total area under the curve lies between $x = \bar{x} - \sigma$ and $x = \bar{x} + \sigma$, as indicated in the figure;

95% of the total area under the curve lies between $x = \bar{x} - 2\sigma$ and $x = \bar{x} + 2\sigma$;

This means that if we make a random observation, there is a 68% chance of the value lying within one standard deviation of the mean and a 95% chance of it lying within two standard deviations.

Viewed the other way round, there is only a 5% chance that we will observe a value more than 2 standard deviations away from the mean.

Exercise

A large set of measurements of the concentration of ozone in urban air is collected and found to have a mean of 4 ppm and a standard deviation of 0.5 ppm. What is the probability of recording a concentration in excess of 5 ppm?

(Answer: 2.5% — we are concerned with a value more than 2 standard deviations away from the mean: there is a 5% chance of observing such values, but half of these would be more than 2σ lower than the mean, the other half, 2.5%, are 2σ or more greater than the mean.)

Viewed another way, if we do record a concentration in excess of 5 ppm, then there is only a 2.5% chance that this value has arisen as a random variation from the mean so there is a greater chance that this indicates that something else is causing the abnormally high concentration on this occasion (e.g. a heavy traffic load).

The greater the number of standard deviations away from the mean that a given reading is, the much less likely it is to arise as a simple random variation.

6.4 Calculating the variance and standard deviation for a finite data set

In practice we are unlikely to be faced with infinitely large data sets. We can, however, evaluate the variance and standard deviation for any size of data set, although these quantities really only have much significance for large sets.

Once the mean value $\bar{x}$ of a data set has been found, then the difference or *deviation* δ_i of any particular reading x_i from this mean value is simply given by

$$\delta_i = x_i - \bar{x}$$

We can evaluate the deviations for every member x_i of the data set to give δ_i for $i = 1$ to n if there are n readings. One possible measure of the spread of the reading about the mean $\bar{x}$ might then be to take the average value of these deviations. We would do this by adding the δ_is and then dividing by the number of readings.

Exercise

Calculate the deviations δ_i for the Cu^{2+} concentration data given in section 6.1 about the mean value (use $\left[Cu^{2+}\right] = 18\,ppm$ and $n = 7$)

From this find $\displaystyle\sum_{i=1}^{n} \delta_i =$

and then the mean deviation $\displaystyle\bar{\delta} = \frac{\sum_{i=1}^{n} \delta_i}{n} =$

(Answer: $\delta_1 = 17 - 18 = -1$, $\delta_2 = 6$, $\delta_3 = 1$, $\delta_4 = -6$, $\delta_5 = 3$, $\delta_6 = -1$, $\delta_7 = -2$; $\Sigma\delta_i = 0$ and so $\bar{\delta} = 0$.)

This final result is not a coincidence. The definition of the mean actually ensures that sum of the deviations is exactly equal to zero for any data set. There are always both positive and negative deviations and these will always exactly cancel each other out.

The real problem is that some δ_is are negative, but it is really their *magnitude* that is important, so a deviation of -2 corresponds to a datum point that differs from (and so is 'as far away as') the mean by just as much as a point with deviation $+2$.

One way round this is to square all the deviations before we sum them, so that they all become positive numbers. We can then take the mean value of this 'sum of squares'. It is this that is called the *variance*, which is given the symbol σ^2

$$\sigma^2 = \frac{\sum_{i=1}^{n} (\delta_i)^2}{n}$$

The *standard deviation* σ is then the square root of the variance

$$\sigma = \sqrt{\frac{\sum_{i=1}^{n} (\delta_i)^2}{n}}$$

Notice that in these formulae, we evaluate the individual δ_is, square them individually and *then* add them.

Exercise

Calculate the $(\delta_i)^2$ values for the $[Cu^{2+}]$ data set

Now, find the variance and standard deviation using the above formulae

$$\sigma^2 =$$

$$\sigma =$$

(Answers: $\delta_1^2 = 1$, $\delta_2^2 = 36$, $\delta_3^2 = 1$, $\delta_4^2 = 36$, $\delta_5^2 = 9$, $\delta_6^2 = 1$, $\delta_7^2 = 4$; $\Sigma(\delta_i^2) = 88$ so $\sigma^2 = 88/7 = 12.6$ and $\sigma = 3.5$.)

Exercise

The mean Cu^{2+} concentration has been found to be 18 ppm with a standard deviation of 3.5 ppm. What is the concentration range within which you would expect to find 95% of any further readings from the same source?

(Answer: we expect 95% to lie within two standard deviations, i.e. within ±7 ppm, thus the range is 11 ppm $\leq [Cu^{2+}] \leq 25$ ppm.)

6.5   Allowing for small sample sizes

The above formulae are really only appropriate if we have 'large' data sets. If we have only a few readings, it is unlikely that these will truly reflect the real frequency distribution curve and so the mean value that we obtain may well not be a very accurate approximation to the 'real' mean of the system. This will then lead to errors in σ^2 and σ.

We can go some way to allowing for this by modifying the formulae above using the *Bessel correction*, which uses $n - 1$ instead of n when taking the average of δ_i^2: so we use

$$\sigma^2 = \frac{\sum_{i=1}^{n} (\delta_i)^2}{n-1} \qquad\qquad \sigma = \sqrt{\frac{\sum_{i=1}^{n} (\delta_i)^2}{n-1}}$$

Exercise

Recalculate the variance and standard deviation for the $[Cu^{2+}]$ data set on the basis of the Bessel correction (remember $n = 7$ and $\Sigma(\delta_i)^2 = 88$)

What is the corrected concentration range within which we can expect 95% of future readings to lie?

(Answer: $\sigma^2 = 88/6 = 14.7$, $\sigma = 3.8$; 10.4 ppm $\leq [Cu^{2+}] \leq 25.6$ ppm.)

We have to accept a slightly larger range and larger uncertainties because our calculations are based on a small sample. Clearly, however, if n becomes large, the difference between using n and $n-1$ becomes less significant and we approach the 'ideal' formulae given earlier.

There is one other quantity that now becomes of interest. As mentioned above, because only a finite number of readings are available, and these are subject to random error we cannot be sure that the mean value that we calculate is actually correct, i.e. is the answer we would get if we had collected an infinite number of data points. If we took another 7 readings, that set might give a slightly different mean concentration.

Can we use information collected from one just data set to assess how much the mean is likely to vary from one data set to another ?

For this we make use of the *standard error of the mean* ε. This is calculated from the standard deviation σ and the sample size n:

$$\varepsilon = \frac{\sigma}{\sqrt{n}}$$

The basis for this is that we assume that if we took a large number of data sets and calculated the mean for each of these, then these mean values would be distributed about the real mean according to a normal distribution.

We use the above equation to express the precision with which we have determined the mean value of the quantity from a finite set of data.

Exercise

Calculate the standard error of the mean for the $[Cu^{2+}]$ data set.

(Answer: we have $\sigma = 3.9$ and $n = 7$, so $\varepsilon = 1.5$.)

This result means that we can be 68% confident that the true mean lies between 16.5 and 19.5 ppm, i.e. between $\overline{[Cu^{2+}]} - \varepsilon$ and $\overline{[Cu^{2+}]} + \varepsilon$. Similarly we would now be 95% sure that the true mean lies within two standard errors of the observed mean, i.e. between 15 and 21 ppm.

It is common to report results with the observed value $\pm$ two standard errors in the following form

$$\overline{[Cu^{2+}]} = 18(\pm 3)\, ppm$$

6.6 Linear regression: the 'best' straight line graph

Data collected in a real experiment are inevitably subject to the random errors described above. If we collect two sets of measurements, perhaps of concentration as a function of time, then we must expect some 'scatter' of the data points about any line or curve on a graph that we subsequently plot.

As mentioned in section 3.7, we most often wish to use the appropriate straight line plot.

For real data, however, even if we choose the correct functions to plot, we will not expect all of the points lie exactly on a line: in fact, we should expect that all will lie more or less 'off' any straight line we draw.

If the experimental points do not lie perfectly on a straight line, then there will be inevitably some choice as to exactly which line to draw 'through' the data set. This will then lead to some uncertainty or error in the gradient and intercept of the line.

Rather than relying on finding the 'best fitting' line by eye, it is conventional to use the method of *least squares analysis* to find the 'best' straight line for a given set of points.

There are many computer-based packages available that perform these linear regression analyses and the need to calculate regression coefficients by hand has mainly ended. Nevertheless, the basic background can be covered relatively quickly and provides some useful insights.

This method is illustrated with the following graph:

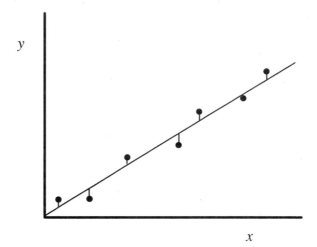

Once a line has been drawn on a graph, we can measure the 'distance' of each point from the line: this 'distance' is defined as being the vertical distance between a point and the point on the line directly above it, as indicated in the figure.

These 'distances' $y_{point} - y_{line}$ will either be positive or negative, depending on whether the point lies above or below the line.

However, the **squares** of these distances $\left(y_{point} - y_{line}\right)^2$ will all be positive.

The method of 'least squares' works by choosing the line with gradient and intercept that makes the sum for these squares over all of the points as small as possible,

i.e.   so $\displaystyle\sum_{all\ points}\left(y_{point} - y_{line}\right)^2$ is a minimum.

This requirement specifies a unique *line of regression*.

In addition, the method allows the calculation of the uncertainty ('error' or 'variance') in the gradient and intercept.

We seek to find the 'best' straight-line fit $y = mx + c$ to a set of data points (x_i, y_i) with $i = 1$ to n, with the 'best' fit being that which minimises the 'sum of squares'

$$\sum_i \left(y_i - y_{\text{calc}}\right)^2 = \sum_i \left(y_i - mx_i - c\right)^2$$

This will be achieved if we calculate m and c from the following formulae:

$$m = \frac{n\sum x_i y_i - \sum x_i \sum y_i}{n\sum \left(x_i^2\right) - \left(\sum x_i\right)^2}$$

and

$$c = \frac{\sum x_i \sum x_i y_i - \sum \left(x_i^2\right)\sum y_i}{\left(\sum x_i\right)^2 - n\sum \left(x_i^2\right)}$$

If we also calculate the 'variance of y' σ^2 from

$$\sigma^2 = \frac{\sum (y_i - mx_i - c)^2}{n-2}$$

then the variances in m and c can be calculated as:

$$\sigma_m^2 = \frac{n\sigma^2}{n\sum \left(x_i^2\right) - \left(\sum x_i\right)^2}, \qquad \sigma_c^2 = \frac{\sigma^2 \sum \left(x_i^2\right)}{n\sum \left(x_i^2\right) - \left(\sum x_i\right)^2}$$

The 'errors' or uncertainties in the gradient and intercept are then usually quoted as $\pm\sigma_m$ and $\pm \sigma_c$ respectively.

The factor $n - 2$ in the variance indicates that more than two points are needed for such a linear regression fit. (Essentially, two are required to find the slope and intercept and we need others to test the 'goodness' of this fit.)

Example/Exercise

The rate constant k for the hydrolysis of 2-methyl 2-chloropropane is found to have the following values at seven different temperatures

experiment, i	T/K	k/s^{-1}
1	293	5.26×10^{-3}
2	298	7.56×10^{-3}
3	303	14.4×10^{-3}
4	308	22.7×10^{-3}
5	313	48.3×10^{-3}
6	318	62.1×10^{-3}
7	323	119×10^{-3}

We wish to make a linear plot of $\ln(k)$ as y against $1/T$ as x to find the activation energy E and pre-exponential factor A from the gradient and intercept.

By completing the following table, we will be able to calculate the best fit slope and intercept using the above formulae.

The first column simply repeats the experiment number or 'index'. The next two columns have the $1/T$ and $\ln(k)$ data to be used for the x and y axes which have to be calculated from the T and k data in the previous table. The final two columns in the table are headed x_i^2 and $x_i y_i$. These are the values of x^2 and $x \times y$ for each experiment.

i	$x_i = K/T$	$y_i = \ln(k)$	x_i^2	$x_i y_i$
1	3.413×10^{-3}	-5.248	1.165×10^{-5}	-17.91×10^{-3}
2				
3				
4	3.247×10^{-3}	-3.785	1.054×10^{-5}	-12.29×10^{-3}
5	3.195×10^{-3}	-3.030	1.021×10^{-5}	-9.68×10^{-3}
6	3.145×10^{-3}	-2.779	0.989×10^{-5}	-8.74×10^{-3}
7	3.096×10^{-3}	-2.129	0.959×10^{-5}	-6.59×10^{-3}
	$\Sigma x_i =$	$\Sigma y_i =$	$\Sigma(x_i^2) =$	$\Sigma x_i y_i =$
$n = 7$	$(\Sigma x_i)^2 =$			

At the bottom of each column we need to find the sums of all the numbers in that column to give Σx_i, Σy_i, $\Sigma(x_i^2)$ and $\Sigma(x_i y_i)$ respectively. There is also an entry for the value of $(\Sigma x_i)^2$. Care needs to be taken to distinguish between $\Sigma(x_i^2)$, which is the sum of the squares x_i^2 and $(\Sigma x_i)^2$ which is the square of the sum of x_i, i.e. we do the squaring and adding in a different order for these two different quantities.

(Answers: $\Sigma x_i = 22.752 \times 10^{-3}$; $\Sigma y_i = -26.097$; $\Sigma(x_i^2) = 7.403 \times 10^{-5}$; $\Sigma x_i y_i = -85.6 \times 10^{-3}$; $(\Sigma x_i)^2 = (22.752 \times 10^{-3})^2 = 517.65 \times 10^{-6}$.

Plot the points as y against x (i.e. $\ln(k)$ versus $1/T$) on the graph below. Estimate the best fit line through the points and determine the gradient of this line.

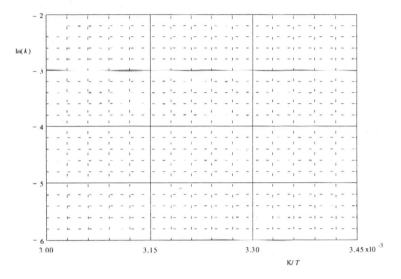

We can now use the formulae given earlier to calculate the gradient m and intercept c of the line of least squares

$$m =$$

$$c =$$

(Answers; $m = -9716$, $c = 27.88$. These results are actually rather sensitive to the number of significant figures retained in each summation; a commercial least squares fitting package gave $m = -9900$ and $c = 28.7$ for comparison.)

This regression line is shown with the data points in the graph below

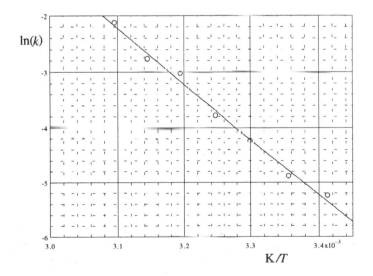

To evaluate the variance of the gradient and intercept values, we need to compare the actual y_i values with those computed from the least squares fit,

$$y_{i,\text{calc}} = m\,x_i + c$$

Enter these data in the table below

i	y_i	$y_{i,\text{calc}}$	$y_i - y_{i,\text{calc}}$	$(y_i - y_{i,\text{calc}})^2$
1	−5.248			
2	−4.885			
3	−4.241	−4.186	−0.055	3.03×10^{-3}
4	−3.785	−3.665	−0.12	14.4×10^{-3}
5	−3.030	−3.162	0.132	17.4×10^{-3}
6	−2.779	−2.673	−0.106	11.2×10^{-3}
7	−2.129	−2.200	0.071	5.0×10^{-3}
			$\Sigma =$	$\Sigma = 78.0\times10^{-3}$

Using the sum of the final column, calculate the variance σ of y and the 'errors' in m and c, σ_m and σ_c

$$\sigma^2 =$$

$$\sigma_m =$$

$$\sigma_c =$$

(Answers: $\sigma^2 = 0.0156$, $\sigma_m = 400$, $\sigma_c = 1.4$)

Using these results we can express the gradient, intercept and their uncertainties in the following form

$$m = -9700\,(\pm 400)$$
$$c = 27.9 \pm (1.4)$$

We can now use these results to find the activation energy and the pre-exponential factor and *also* the experimental uncertainties in these quantities

Exercise

Use the results for m and c along with the relationships
$$m = -E/R$$
and
$$c = \ln(A)$$
where $R = 8.314$ J K^{-1} mol^{-1} to find E and A.

(Answers: $E = -R \times m = 80.6\,(\pm 3.3)$ kJ mol^{-1}; here we have taken the uncertainty to be $\pm$ one standard deviation, i.e. $\pm 400 \times R$, 3.3 is the same percentage of 80.6 as 400 is of 9700: $A = e^{27.9\pm1.4}$, taking the number without the uncertainty gives $A = 1.3 \times 10^{12}$ s^{-1}; if we re-evaluate A with the maximum and minimum values 29.3 and 26.5 in the exponent, we get $A_{\text{max}} = 5.3 \times 10^{12}$ s^{-1} and $A_{\text{min}} = 3.2 \times 10^{11}$ s^{-1}, indicating that the 5% uncertainty in $\ln(A)$ leads to an uncertainty of a factor of approximately 3 in A itself due to the exponential dependence of A on the value of the intercept.)

Summary of Section

The material in this section has been concerned with the basics of statistics in the way we need to apply them to experimental data in chemistry. The following concepts should be familiar:

- the idea of a distribution of repeated measurements of the same quantity

- the idea of a mean value as representative of a sample $\bar{x} = \dfrac{\sum\limits_{i=1}^{n} x_i}{n}$

- the shape of the normal distribution curve and the position of the average
- the idea of a standard deviation and its relationship to the width of the distribution curve

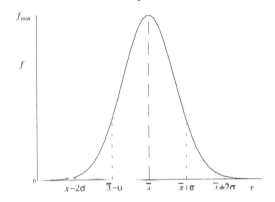

- the recipe for calculating the variance and standard distribution for a set of measurements $\sigma = \sqrt{\dfrac{\sum\limits_{i=1}^{n} (\delta_i)^2}{n}}$

- how to allow for small sample sizes (the Bessel correction) $\quad \sigma = \sqrt{\dfrac{\sum\limits_{i=1}^{n} (\delta_i)^2}{n-1}}$

- the standard error of the mean $\quad \varepsilon = \dfrac{\sigma}{\sqrt{n}}$

- how to find the 'best' straight line passing through a set of real data points and how to calculate its gradient and intercept and the errors in these

$$m = \frac{n \sum x_i y_i - \sum x_i \sum y_i}{n \sum (x_i^2) - (\sum x_i)^2} \qquad c = \frac{\sum x_i \sum x_i y_i - \sum (x_i^2) \sum y_i}{(\sum x_i)^2 - n \sum (x_i^2)}$$

$$\sigma^2 = \frac{\sum (y_i - mx_i - c)}{n-2} \quad \sigma_m^2 = \frac{n \sigma^2}{n \sum (x_i^2) - (\sum x_i)^2}, \quad \sigma_c^2 = \frac{\sigma^2 \sum (x_i^2)}{n \sum (x_i^2) - (\sum x_i)^2}$$